DATE			

GRAVITY'S LENS

GRAVITY'S LENS

VIEWS OF THE NEW COSMOLOGY

Nathan Cohen, Ph.D.

Wiley Science Editions
JOHN WILEY & SONS, INC.
New York · Chichester · Brisbane · Toronto · Singapore

THE WILEY SCIENCE EDITIONS

The Search for Extraterrestrial Intelligence, by Thomas R. McDonough
Seven Ideas That Shook the Universe, by Bryon D. Anderson and Nathan Spielberg
The Naturalist's Year, by Scott Camazine
The Urban Naturalist, by Steven D. Garber
Space: The Next Twenty-Five Years, by Thomas R. McDonough
The Body in Time, by Kenneth Jon Rose
Clouds in a Glass of Beer, by Craig Bohren
The Complete Book of Holograms, by Joseph Kasper and Steven Feller
The Scientific Companion, by Cesare Emiliani
Starsailing, by Louis Friedman
Mirror Matter, by Robert Forward and Joel Davis
Gravity's Lens, by Nathan Cohen

Publisher: Steven Kippur
Editor: David Sobel
Managing Editor: Ruth Greif
Editing, Design, and Production: G&H SOHO, Ltd.
Illustrator: Charles Messing

Library of Congress Cataloging-in-Publication Data

Cohen, Nathan.
Gravity's lens.

(The Wiley science editions)

Includes index.
1. Cosmology. 2. Gravitational lenses. 3. Radio astronomy. I. Title. II. Series.
QB981.C64 1988 523.1 88-14353
ISBN 0-471-63282-1

Printed in the United States of America

88 89 10 9 8 7 6 5 4 3 2 1

"We've looked and looked, but after all where are we?
Do we know any better where we are,
And how it stands between the night tonight
And a man with a smoky lantern chimney?
How different from the way it ever stood?"

—from "The Star-Splitter" by Robert Frost

Foreword

THE history of science shares many a common theme with all other histories; for example, that events will be so often controlled by human altruism, or, too often, human greed. But there is an oft-repeated theme special to science history, and one that should humble us: the notion that we have finally known everything important. That notion has reared its head so many times, often with the unfortunate side effect of discouraging talented young people from embarking on scientific careers. Each time it has been dashed by the next important and unexpected great discovery, bringing joy and excitement to the scientists of the time, and renewed prospects for those contemplating scientific careers.

If there is a single message in this book, it is that once again we are in an era of great, revolutionary astronomical discoveries. That, again, the supposedly great truths were far from the full story. Pouring from this book is a cascade of important observations and theories, almost all of them products of the last ten years. So much do they alter our views of the structure and history of the universe that we had best do our reading in small doses. It may be uncomfortable to see too many old, cherished ideas dismissed in one reading session.

In the 1920s we knew it all; then we discovered to our surprise that there was more than one galaxy; indeed, the sky was full of galaxies. And soon thereafter we found that they were receding from us in a perfectly simple and predictable way, following the laws of Edwin Hubble. What a bombshell! From this grew the profound idea that the universe had not always existed, but had started at a finite time, in a grand explosion, not too long ago, at least by cosmic standards. This result was confirmed by the discovery that no matter where in the sky one pointed a radio telescope, one saw a faint red glow. This perpetual cosmic dawn was the primordial fireball itself, the original explosion seen by us as we looked back through time almost to its beginning. This observational tour de force was properly the basis for a Nobel prize. What had come before the primordial fireball? That is still a question beyond the realm of science.

Then remarkable probing minds noticed something which might seem entirely proper, but was not: the universe was smooth, the galaxies flung almost uniformly across the entire expanse of the universe. How could the chaos of a violent explosion have created so smooth a pattern? Only if there had been an unusual change of state, a remarkable expansion at fantastic speeds, in the first few instants of the existence of the universe.

From this comes some remarkable ideas, all to be found here. The precise nature of all reality, even the existence of life, depends on the design imposed on the universe in those early instants of time. The very building blocks of all matter were created then, along with the physical laws that would govern them. The foundations were laid just then for the existence of stars, of planets, of carbon, and all that we know, from redwood trees to humans to modern computers.

Along the way some remarkable objects were created. The quasars, the gargantuan destroyers of we know not what. The blazing light that is the debris of these conflagrations is seen clear across the universe. The black holes? Surely they are there somewhere; read on, dear reader. And perhaps the strangest things of all: other universes! "Clones" perhaps of our own, although it is possible, even probable, that in them the physical laws are entirely different from here. Perhaps in those places there never will be stars, nor planets, nor even a sentient being to wonder how it all came about. Only darkness. Or perhaps there will be but a single star in all the sky, the object of wonder to creatures so strange that we could never image them. It is possible. But we probably will never know; universes are, we think, forever isolated from one another, unable to send across space from one to the other even the simplest words of greeting.

We are destined to know much more about our universe. Out there in space, waiting to be used, are the most powerful lenses of all: lenses created by the gravitational bending of light. Their great potential was recognized so many years ago by Einstein, yet only in the recent past have we witnessed their remarkable power. Rather than bringing rays of light to a focus through the bending of the rays in glass, or reflection from a suitable mirror, they actually focus light through the substantial bending of space that occurs in the vicinity of any massive object in our universe. As a result, if the space bending object is a galaxy, the light from a zone whose dimensions are of the order of the size of a galaxy is brought to a focus. If the bender is a star, the zone is as large as a star.

Clearly, such gigantic lenses can provide magnification and image sharpness that far surpass the abilities of even our most heroic telescope designs. Indeed, the galactic benders can create bright images of faint galaxies at enormous distances. The gravitational lenses made by stars can create sharp images of stars and planetary systems across distances of hundreds or even thousands of light years. They may even amplify the radio transmissions of distant worlds. Due to the enormous amplification of stellar gravitational lenses, even our modest human transmissions echo to the farthest stars! We are just beginning to recognize the power of this phenomenon; we have not yet begun to utilize it. Its full utilization will require abilities in space technology that are just beyond our grasp; it is a project perhaps for the twenty-first century. When exploited, the gravitational lenses will project before us a clear picture of the universe in all its grandeur.

Nathan Cohen has been one of the small group of early observers of this phenomenon, and he describes it extremely well here. He describes how galaxies serve as giant lenses, giving us clear images of faint objects at the far side of space. He shows how the gravitational lens opens that powerful new window into space, a window that will certainly reveal, as did our present telescopes, an unexpected cosmic wonderland.

The invention of each new astronomical instrument has produced a wealth of new discoveries. Each time these discoveries have reminded us that we live in a universe so diverse in its phenomena that only rarely do we predict the remarkable objects and activities there. In this book we see this once again, through its description of the discoveries and new ideas that have so remarkably changed, in just a few years, our concepts of the structure and history of the universe. We also see clearly that the thread of scientific history is intact: the greatest discoveries are yet to come.

Frank D. Drake
Professor of Astronomy and Astrophysics
Dean, Division of Natural Sciences
University of California, Santa Cruz

Preface

WHILE the idea for this book was planted some time ago, it had to await a modest but tangible time span before seeing reality. But I'm glad I waited. For these are heady times in observational cosmology, full of progress and vitality. The writing of this book was an especially satisfying excuse to keep up with a very fast-moving science in which few insiders get a chance to take a step back and get the big picture. One feels a slight guilt in putting research aside, even temporarily, but the experience of writing has great compensations, too: There is a great joy in knowing, but that joy is multiplied in the telling. And the universe is a subject all should be able to find out about, even if a nontechnical but driven curiosity is the principal motivation.

Still, this book has proved a great challenge, in part because the universe reveals itself only in the most mysterious of ways. Only a decade ago, cosmology seemed almost dull; the CBR had been found, Hubble's law was hoary and matter of fact, and everyone knew that the universe was open—or so it seemed. Cosmology was an interesting science but not necessarily one rife with continual breakthroughs. But by 1980 or so, tantalizing things were coming to light: the voids, a gravitational lens, hints of ubiquitous dark matter, the SS 433 jet-star, and more. Streaming, a nearby supernova, high redshift galaxies, and the inflation model were either just emerging or yet to be believed or discovered. Large-scale structure was a seldom-used buzzword. Galaxies at earlier times were a great unknown, and particle physics was only beginning to demand attention as a prime factor in contemporary cosmology. By the early years of the decade, cosmology was emerging as a science not yet willing to accept a mostly historical connotation.

Then again, the history of the field suffered from a few myths. Imagine my surprise to find that the humble but brilliant physicist H. P. Robertson was the discoverer of Hubble's law and, arguably, of the expanding universe as well (the famous cosmologist Allan Sandage has independently revealed this point in his recent essay in the book *The Universe*). Or that Fred Hoyle, inspired cosmologist and science writer, was the first to come up with the largest angular-size redshift test. Or that maverick astronomer Fritz Zwicky predicted a gravitational lens effect for galaxies decades before one was seen.

My hope for the reader is that he or she will see what a fascinating puzzle the universe is, demanding ever-changing strides in theory and observational technique and hardware. To assist the reader on this journey, I have made it a point of using a copious number of pictures and diagrams to illustrate some of the less-than-common sense ideas that come up in cosmology. One can easily argue the aesthetic merits of some of these illustrations; indeed, many have a distinctly abstract quality—is nature imitating art? One should be cautioned that many of these illustrations have been produced through computer-enhanced image processing and are false-color representations. Color is used as an aid to help the eye distinguish details and is not necessarily a representation of real color shading (for example, what would the "real" colors be in a radio "photograph"?). And in some diagrams, contouring is used in place of false-color shading. Astronomers are fortunate in being able to show works of art as evidence for profound scientific points!

An author may be the sole writer of his story, but he does not function in a vacuum. The project would have been a hollow experience without the aid of my colleagues, who provided clarification, information, and/or material. Succinct but sincere thanks to these individuals, who include Chip Arp, Barry Geldzahler, Norbert Bartel, Martha Hazen, Martha Haynes, Matthew Malkan, Chris O'Dea, Len Cowie, Emilio Falco, Jean Surdej, Ed Krupp, Joan Centrella, Brent Tully, Roger Angel, Ed Loh, John Tonry, Jim Peebles, Alan Bridle, Phil Appleton, Vera Rubin, François Schweizer, Marc Davis, Arno Penzias, Paul Feldman, Rudy Schild, John Biretta, Roger Lynds, Alan Guth, Sidney Van den Bergh, and Susan Simkin. I was especially grateful for a chance to raid several excellent photo collections, including those of the Millikan Library of Caltech, the Caltech Bookstore, the Mount Wilson Observatory, NRAO,

NOAO, and the AIP. And Charles Messing's faithful renditions of the diagrams added immeasurably to their clarity and impact.

Although this book saw its beginning as I was finishing my doctorate at Cornell, it briefly encountered a school or two along the way before completion at Boston University. I am grateful for the opportunity to complete this book at B.U. and thank my colleagues at MET for creating an encouraging atmosphere for its final stages.

Finally, it is crucial to thank David Sobel and his staff at Wiley for their infinite patience, help, and good judgment, not to mention the chance to present *Gravity's Lens* in just this way.

N.C.
Boston, Massachusetts

Contents

ONE

PRE-VIEWS

· ONE ·

Unveiling the Universe

HOW much can we ultimately find out about the universe? Each successive generation of astronomers has pondered this question, hoping, through observation and insight, to hone in on an answer. And at last, if the progress of the last decade proves a worthy guide, we are beginning to understand the origin, evolution, and fate of the universe.

For this has been a remarkable and exciting decade in the field of cosmology—the science of the universe. Many important discoveries have been made, usurping less-viable theories and less-complete observations, some under a generation old.

Consider just a few of these discoveries:

- Gravitational lenses: galaxies that shape the light from more distant galaxies into bizarre mirages
- Cosmic strings: defects in the geometry of space and time, stretching across billions of kilometers
- Inflationary universe theory: an explanation for our origin that holds that the universe began from an unstable vacuum, expanding from "nothing" to "something" almost instantaneously
- Bubbles and voids: balloonlike groups of galaxies that span the universe as superclusters.

A mere decade ago, such odd findings would have been considered fancies of science fiction rather than mainstream science.

This era of cosmic exploration is set apart from earlier ones by the surprising nature of these and other findings. This is truly a time of a new cosmology: Our picture of what the universe is, was, and will be is more detailed now than ever before. By exploring these new findings, we attain a better grasp of what the universe is like and experience the wonder of its workings.

Every breakthrough era of a science has a foundation on which it is built. In the case of cosmology, this foundation has changed many times, mirroring the progress of our intellectual and technological growth. Yet the important questions have always been the same—those questions of inquiry into the unknown. Tracing the evolution of cosmology will provide an important and interesting background into the development of where we are now. Here we will consider some of the most important aspects of cosmology through the centuries, a story that begins with the eye and the imagination and ends with the fires of creation.

FROM NAKED EYE TO A DISTANT STAR

We base so much of what we know about the world on what we see for a good reason. Of all the senses, sight is the one that tells us the most information about size, structure, numbers, and situations. And it is the only sense that allows us to perceive the world from a great distance. Yet vision provides only meager clues to the staggeringly large distances of the stars and galaxies.

We cannot blame nature for this, because it has equipped us with vision as a superior tool for sensing our environment. We can use information about our environment to increase our chances for survival. Celestial sights are bonuses. Fortunately, along with the need for survival, intellect and reason evolved. The sky, a realm beyond our immediate environment, was not beyond our curiosity and desire for understanding.

Early man used mythology and guesswork, rather than knowledge and reason, to describe the nature of the sky. Ancient civilizations constructed vast pyramids and temples, many of which seem to have only religious meaning. Others, such as England's Stonehenge, may have served as accurate and elaborate timepieces, heralding in the changing weeks of the year with the shifting positions of the constellations. The sky's first "purpose" then, must have been that of a cosmic clock, foretelling periods of planting and reaping. Astronomy's early coupling with agriculture makes the study of the sky an endeavor as old as civilization. (See Plate 1 following page 78.)

Early civilizations also developed astrology, the belief that human cycles were related to those of the sky. The changing yet predictable patterns of the heavens also inspired a belief in eternal life and a realm of mythical beasts, gods, and goddesses. Although enchanting as fantasies, such notions failed to give concrete answers on the nature of the stars and sky. Yet many of these same ideas still inspire us today, despite our sophisticated knowledge about the universe.

The Greeks were among the first to go beyond superstitious ideas about the sky. Many ancient Greek astronomers possessed keen and methodical observational skills and were able to achieve the maximum results possible under the limitations of the human eye. Combining their visual and analytical skills, Greek astronomers such as Aristarchus and Ptolemy realized that the sun, moon, planets, and stars were very different from the wind and clouds. Based on planetary motions and occasional planetary alignments, they surmised a relative ordering of distances for the planets. And a millennium before other Europeans accepted the concept, the Greeks concluded that Earth was spherical by tracing its shadow during a lunar eclipse.

Among their many other profound contributions is the mathematical study of trigonometry, a sophisticated method for calculating distances that are often impossible to measure physically. To find

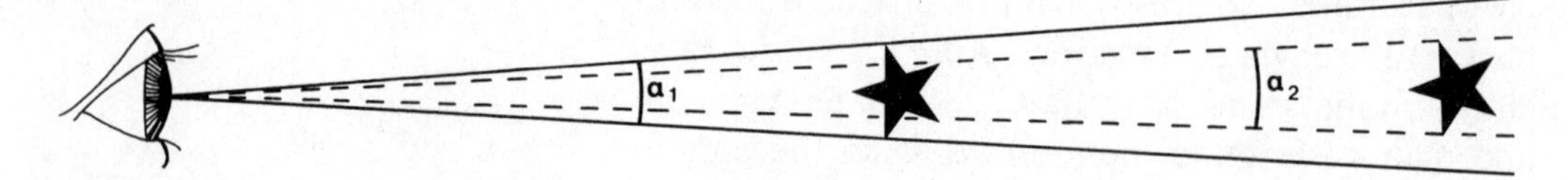

An object's physical size relates to the angle it takes up—its angular size—by its distance. When an object is moved farther away, its angular size decreases, as shown by the two different angular sizes we see for this star.

an unknown distance, all one needs to know are the distances and angles with respect to certain reference points. Trigonometry also established the relation between an object's apparent size and its actual physical size and distance. Simply stated, the *angular size* of an object is how big an angle it takes up. Everyday experience shows that the closer an object is, the larger its angular size; for example, a nearby car has a large angular size, while at a great distance the same car appears much smaller. Trigonometry shows that, for a given value of angular size, the ratio of the physical size to distance is constant. In other words, a large, distant object can have the same angular size as a smaller, closer one. But if we know the physical size of an object, we can discover its distance through trigonometry by measuring its angular size. Additionally, trigonometry can give distances through triangulation or a change in perspective. Trigonometry is a powerful tool for finding distances.

The Greeks used trigonometry with great care and found it useful in finding distances on Earth. But they also realized that the pinpoints of light that appeared as stars and planets were too far away to use the eye's limited facility for finding minute angles. When it comes to these great distances, the eye is depth blind.

Despite depth blindness, the Greeks came up with the first scientific model for the universe. They maintained that Earth was the stationary center of a universe made up of spheres, or shells, and that each shell corresponded to the position and path of a heavenly body. The outermost and most perfect of these shells was the sphere of the stars. The idea makes much sense, for it describes, at a naive level, what we see everyday: The sun, moon, stars, and planets move, while Earth stands still. Even today we find it easy to revert to this false but descriptive idea, and more than a few astronomers commonly use such phrases as "the sun rises" or the "moon sets," even though it is Earth's rotation that gives these effects.

In 1515, nearly a millennium after the Greek astronomers, the Polish astronomer Nikolai Copernicus pursued a different approach, resurrecting a model for the universe that the Greeks had conceived but discarded because it challenged their concept of a perfect universe. To explain the wiggling paths of the planets, Copernicus reversed the ordering of the spheres, making the sun the center of his model and Earth one of the planets. This *heliocentric* model signaled in a revolution of new thought; the European reverence for ancient Greek ideas, which had raised them to the level of dogma, was now toppled by new ideas that worked better in explaining the planet's paths. The real excitement, though, was that Copernicus had shattered the image of Earth as a unique and privileged place. With Copernicus's new ordering of the universe, other scientists were willing to entertain ideas on the distance of heavenly bodies and the makeup of the universe.

One of these scientists was the Italian Galileo Galilei. Renowned for many discoveries in several

Two conventions for the arrangement of the universe as considered by early astronomers: (a) the geocentric model, where Earth is the heart of the universe, and (b) the heliocentric model, where the sun is the center of the planets' motions.

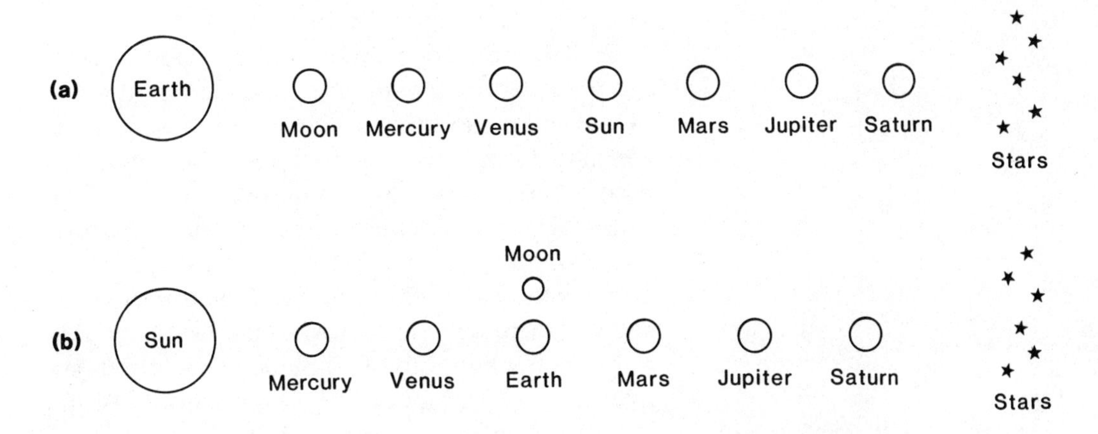

Galileo Galilei, Renaissance astronomer who pioneered the study of the heavens with the telescope. (Courtesy American Institute of Physics, Niels Bohr Library.)

branches of science, Galileo holds the distinction of being the first to apply the telescope to a concerted study of the sky. Earlier, lenses (the earliest of which were polished gemstones) had been restricted to military use and spectacle wear. But in 1610, Galileo combined lenses in a tube to make a refracting telescope of modest but unique magnifying power. With it he hoped to study planetary motions and celestial distances.

Galileo's telescope was a turning point in celestial observation: It radically changed the quantity, detail, and perspective of what could be seen. Feeble light from dim objects could now be viewed with greater clarity, extending the realm of exploration. Eyes alone can see the moon, the sun, six planets, a few meteors, and an occasional comet, along with over 4,000 stars. Yet even a simple telescope can reveal billions and billions of stars, lunar craters and mountains, planetary features, star clusters, and much more. The experience of the unaided eye did not prepare Galileo and other medieval astronomers for these unexpected findings; the telescope opened up a new universe, a new frontier.

Perhaps it is no coincidence that Galileo was a good experimenter as well as a good observer. Indeed, along with other medieval and Renaissance scientists, he refined the concept of an experiment—the mainstay of the scientific method. He developed a system for hypothesizing, predicting, and testing. And more significantly, Galileo displayed objectivity; he was willing to discard ideas that just didn't hold up to experimental testing. Curiously, the willingness to discard incorrect ideas was a bit aberrant in Galileo's time, a time in which church dogma ruled that Copernicus's heliocentric model was disreputable. But Galileo, armed with his new findings of the heavens, was so convinced of the correctness of the heliocentric model that he risked—and received—church censure and court arrest for his views.

MEASURING STELLAR DISTANCES

The Greek astronomers understood the importance of trigonometry, but their lack of telescopes restricted them to the limits of the human eye. Galileo, who took giant steps forward with the telescope and the scientific method, was cut short in his interest in finding out the distance to the stars. But later astronomers were able to use the telescope to look for the minute angles that trigonometry promised would reveal the distances to the stars.

The most important way that trigonometry could be used to determine stellar distances was through the use of *trigonometric parallax,* which we shall call parallax for short. Parallax is the change in the position of a distant object against a backdrop of more-distant ones, caused by a change in the perspective of the observer. An everyday example of a type of parallax involves viewing ob-

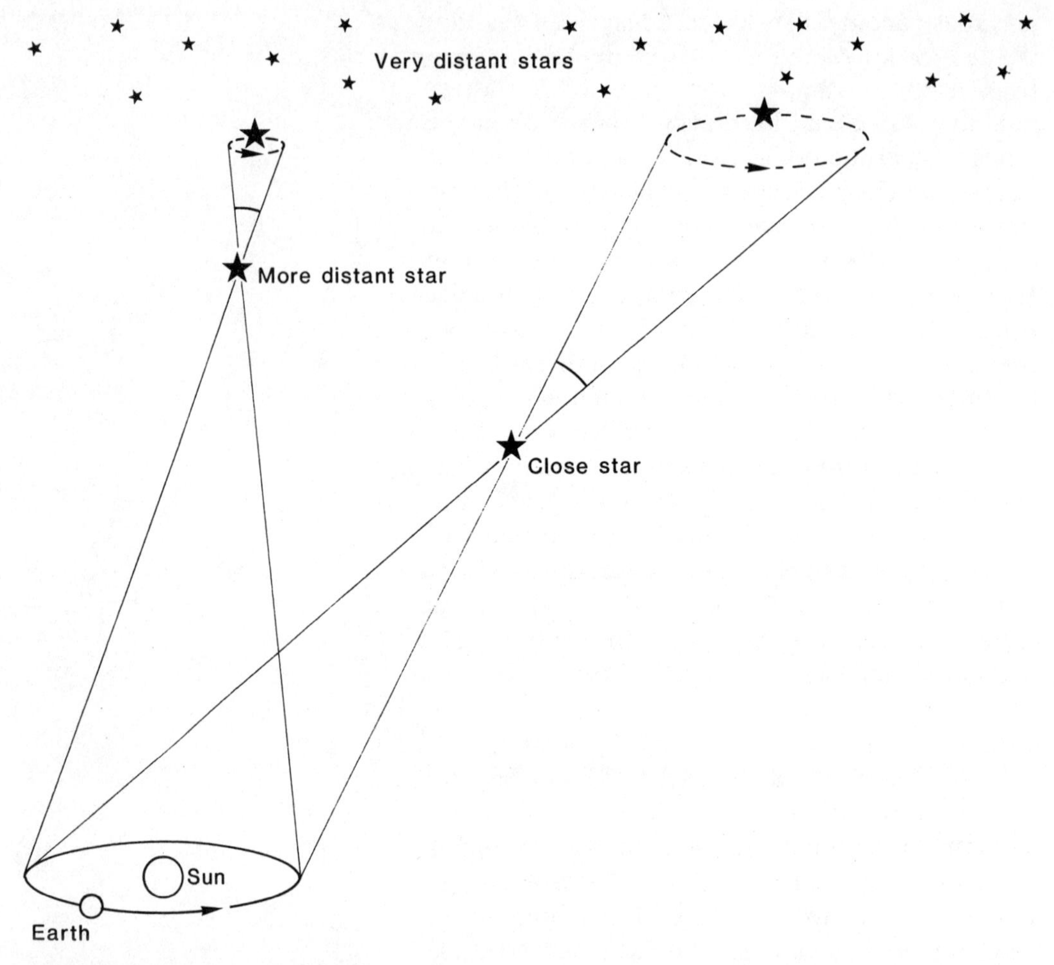

The relative amount of parallactic shift for two stars situated at different distances. The position of the closer star shifts by a larger amount, against the backdrop of the most distant stars, than a more distant one. Parallax is the shifting of the stars' positions, caused by Earth's yearly motion around the sun.

jects, at various distances, from a moving car. As an observer in the car, your perspective of the scene is constantly changing. As a result, nearby objects seem to zoom past while very distant ones appear to move slowly, if at all. In the parallax of stars, Earth—like the car—is moving as it orbits the sun over the course of the year. By comparing the view of stars every six months, you are viewing with a change of perspective, whose distance separation is twice Earth's distance from the sun. Any changes in the position of stars over this period can be used to decipher their distance.

It is easy to use trigonometry to get stellar distances—once a parallactic shift, or *parallactic angle,* is found. Because the angle must be quite small, it is measured in fractions of a degree, in units called *arcseconds;* for example, the moon's angular size is about 1,000 arcseconds. When a star has a parallactic angle of one arcsecond, its distance from Earth is about 3.3 times the distance that light travels in one year; for convenience, astronomers define a unit of distance, the *parsec,* to describe this distance. This convention makes much sense because any star with a parallactic angle of, say, one-third of an arcsecond can then be said to be at a distance of three parsecs.

Finding stellar parallax was impossible for observers without telescopes because the size of the shift is so small that the eye cannot perceive it. Only a telescope, which magnifies the angular shift of parallax, makes it possible for the eye to judge the shift. Even so, the problem is formidable because it entails looking for position shifts of some stars in comparison with a field of others. The task of the stellar parallax hunter was to draw accurate and unbiased charts of many regions of the sky, repeated many times over several years. Yet in the eighteenth and nineteenth centuries, a few astronomers found this task a worthy pursuit.

Friedrich Bessel was one such astronomer. A Prussian businessman, Bessel was a gentleman-scientist in an era when scientific research was a luxury only the wealthy or patronized disciple could afford. Few had the resources, time, and talent to pursue astronomy, which was only beginning to become a respected field eligible for significant university or government support. Although successful in commerce, Bessel became discontent. Early in life, he retired to a life of science, with profound results.

Bessel was a one-man science department who made key discoveries in mathematics, surveying, and other fields. But it was in the field of astronomy that he made his most important contributions. With a penchant for scrupulous, objective observation, Bessel spent four decades investigating stellar parallax. With only a telescope, hand-drawn sky charts, and a keen eye, Bessel mapped the positions of hundreds of stars, searching for the slight, repetitive angular shifts of the closer stars that would indicate parallax. Eventually his patience was rewarded by the discovery of a concrete shift. In 1838, Bessel was able to show that the star 61 Cygnus A had a parallactic angle of 1.5 arcseconds. This angle placed the star at a distance of over 100,000 times the previous recordholder—Saturn. Bessel soon found other stars with similar values of parallactic angle, which implied that the realm of the stars was one that began at an awesome distance and continued to regions still uncharted.

The great reach to 61 Cygnus A was unexpected because there had been no hint that the distance to the stars was so huge. It destroyed an assumption of continuum first stated by the Greeks, and propounded long after the Renaissance, that the stars were as far beyond the planets as the planets were from one another. Now the stars were known to be so distant from the planets that there seemed to be a gap in space. The distance to the stars redefined entirely the scale of the universe.

By the 1850s, the advent of photography brought an end to the eye/telescope era of celestial research. Less than two generations after Bessel's discovery, the first photographic cameras were attached to telescopes and modern astronomy was born.

The camera's main contribution to astronomy is that it allows us to make much more precise observations and measurements than we can achieve with the eye. A photograph gives an exacting, permanent record, free from errors caused by our fallible memory and the fleeting illusionary tricks of the eye. And with the camera it became possible to see millions of stars invisible or dim to the eye. This is because a time exposure picks up dim light, while the eye "refreshes" itself every fraction of a second, erasing the previous view. Using cameras, astronomers could study more stars, more precisely, than Bessel could, placing yet farther distance limits to yet more distant stars.

But even this approach reached a natural limitation caused by the atmosphere. We often think of the cloudless sky as a perfect window. In fact, turbulent air pockets, always present in the upper atmosphere, act as weak and changing lenses, continuously distorting the light of the stars. This gives rise to the twinkling of stars—a very quick variation in their brightness. It also causes the shapes of the images of stars and planets to change, turning a stellar pinpoint of light into a quaking smear. Thus every photograph of a star comes out as a disk rather than a point, making it difficult to measure precisely parallactic angles. It is impractical to detect parallax from stars more than 20 to 30 times the distance of 61 Cygnus A from Earth. Other methods, which do not rely on measuring parallax, are needed to calculate the distances to the vast majority of stars beyond the nearby parallactic ones.

ASTROPHYSICAL DISTANCES

Astrophysics, a relatively new branch of astronomy, offers other ways of finding the distances to the stars. Astrophysics is concerned with investigating the composition, mass, size, structure, and energy of stars, galaxies, and other celestial bodies. In its formative years, at the turn of this century, astrophysicists concentrated on the variable stars—those whose brightness changes with time. There are many different classes of variable stars, from cool (compared with the sun) but gigantic long-period variables to eruptive variables—those with violent, transient bursts. One special type of variable star is the Cepheid variable, named after the constellation in which the first was discovered. With the preciseness of measurement afforded by

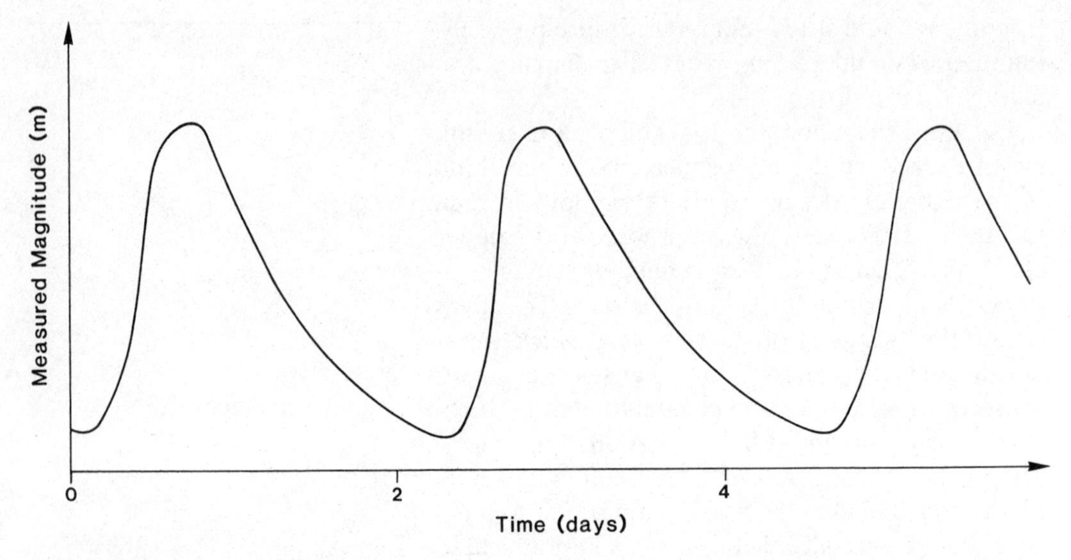

The light curve—a plot of measured magnitude and time. Here, the time of repeating variation of a Cepheid's peak magnitude can be used to infer its absolute magnitude. Comparison with the measured magnitude allows the distance to be inferred.

photography, astronomers found that the brightness of a Cepheid varied in a rapid (time scale of days), repetitive way and that its intrinsic, or *absolute,* brightness was tied into the time it took for the brightness to vary back to its brightest point. Cepheids become dimmer and brighter with a periodic regularity, and each different rhythm is indicative of a specific value of intrinsic brightness.

The brightness change in Cepheids is caused by a tug of war between gravity and the pressure caused by the star's light. Gravity pulls the outer layers of the star in at the same time that light pressure pushes it out. This makes the outer parts pulsate, much like a balloon made bigger and smaller with increasing and decreasing inflation. The process of pulsation is of interest in itself, but its important spin-off is that the time between pulsations will depend on how strong the light pressure is, and thus how intrinsically bright the star is. So the time scale, or period, of the pulsation is a good indication of the star's absolute brightness.

When we measure the brightness of any star, experience tells us that its value will depend on how bright the star is in an absolute sense, as well as how far away it is. Scientists have long known that one of the factors of the brightness of light is the *inverse square law.* Simply stated, the law dictates that every time the distance to a light is doubled, its measured brightness goes down to one-fourth its previous value. We can prove this with a candle. Placed at varying distances, it will be one-fourth as bright at double the distance of its initial placement, one-sixteenth as bright at four times the distance, and so on. But eyes and photographic film share a curious attribute: They do not register brightness changes in terms of factors of 1, 2, 3, 4, and so on. Instead, brightness is described by a logarithmic scale called the *magnitude* scale, where one magnitude of brightness change equals a difference of about a 2.5 factor. The brightest stars are measured at about zero magnitude, while dimmer ones are assigned larger magnitude values. On a crystal clear night, the eye can barely see the dim stars of 5th magnitude while just a simple camera might reveal 9th or 10th magnitude stars—over 10,000 times dimmer than the brightest ones. Some of these dim stars are more distant than the bright ones, while others are just not giving off as much light and are relatively nearby. Clearly, the distance and absolute brightness of a star both contribute to the magnitude we see.

With brightness measured on the magnitude scale, the inverse square law needs to be changed to account for magnitude's logarithmic nature. But the value of this change is great—distance can be found from the new magnitudes. Specifically, if the

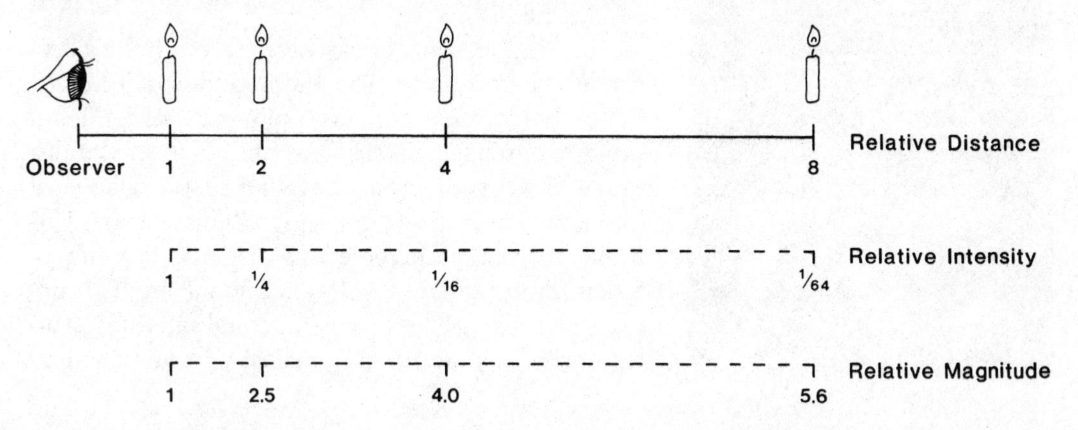

A candle placed at different distances has different measured brightnesses and different magnitudes. The logarithmic magnitude scale uses increasing values for greater dimness.

measured magnitude (m) and the *absolute magnitude (M)* of a celestial object are known, a relation called the *distance modulus* tells how to get the distance. With a distance (D) estimated in parsecs, the distance modulus states that $m - M = 5 \log D - 5$. Brightness becomes a powerful key to estimating the reach to the stars.

With the distance modulus, astronomers could take the period of a Cepheid's variation and, using it as an indication of the absolute magnitude, compare it with the measured magnitude to get the distance. This is a very different way of getting distances, totally unlike parallax, which depends on angular shifts.

Cepheid variables promised a giant step forward in getting stellar distances. In 1912, American astronomer Henrietta Leavitt, discoverer of this period-magnitude relation for Cepheids, used her discovery and the distance modulus to explore the distance to Cepheids in some peculiar nebulae—freely shaped groupings of stars and gas. She concentrated on two nebulae of the southern skies, those called the Large and Small Magellanic clouds (first reported by Magellan after one of his round-the-world voyages). She found many Cepheids in these nebulae, Cepheids that had an unusual characteristic—very faint measured magnitudes. By attributing their faintness to their great distance, Leavitt showed that the Magellanic clouds must be at least a thousand times more distant than 61 Cygnus A.

Harlow Shapley, twentieth-century astronomer and pioneer cosmologist. (Courtesy Harvard College Observatory.)

This finding sent shock waves through the astronomical community. The Large and Small Magellanic clouds were not mere pinpoints of light but great groupings of stars, gas, and dust. At the distance Leavitt inferred, they had to be billions of kilometers across and contain millions of stars. This implied that the universe was made up not only of a continuum of stars, with occasional small groupings, but also of large stellar conglomerates. This fact prompted additional curiosity in nebulae and

Globular cluster M13, a swarm of over 50,000 stars. Shapley inferred the distance to globular clusters using Cepheids as a distance yardstick. He believed globular clusters represented the edge of the universe. (Courtesy Palomar Observatory.)

the related star clusters, particularly with regard to their distances.

At about the same time as Leavitt's work, Harlow Shapley, an American astronomer with farm-boy roots, began a study of star clusters. He hoped to use the Cepheids as a yardstick for distances to the clusters. Specifically, he was interested in the globular clusters: star clusters of spherical shape with up to 100,000 stars—veritable beehives of stars. Globular clusters contain relatively few variable stars, but Shapley found it possible to identify at least a handful of Cepheids in each globular cluster he studied. By looking at several dozen globular clusters, Shapley hoped to find their distances and learn something about the way they were placed in the universe.

What Shapley found came as yet another surprise to the astronomical community. Globular clusters are never found nearby; they are all distant, much more so than the stars that show parallax. Using their distances and positions on the sky, Shapley charted a three-dimensional map of the distribution of these clusters. He found them to be spaced like the seeds on a dandelion head. The most distant was about 100,000 parsecs, or 100 kiloparsecs, away. Shapley concluded that the universe was a realm of stars—what we call the Milky Way—bordered by the globular clusters. In effect, Shapley inferred that the universe had a finite size and was made up of one huge, complex mass with the globular clusters at its edge.

Actually, Shapley was addressing, through his observations, a question best defined by such nineteenth-century Europeans as the philosopher Immanuel Kant and the astronomer John Herschel: How large is the universe? In essence, there are two possible answers: (1) The universe is finite (limited in scale) or (2) the universe is infinitely large (it goes on forever). Proponents of the second answer believed that the nebulae were island universes, of which the Milky Way was but one, and not the whole universe itself.

But in 1826 the Swiss astronomer Heinrich Olbers had proposed an interesting problem whose outcome supported the first theory. Now called Olbers' paradox, this problem involves a clever chain of logic based on an obvious fact—the sky is dark at night. According to Olbers, if the universe were infinite (and infinitely old), then in every direction you looked you would see a star. Even though some of those stars would be very faint, the number of stars in each direction would increase rapidly with distance. As a result, the faintness of each star would be balanced out by the increasing numbers. The sky should look like one continuous blaze of light, rather than a mixture of stars and darkness. Olbers surmised that the real sky suggested that the universe was not infinite.

Shapley's model for the universe was consistent with the outcome of Olbers' paradox, but was the model correct? It certainly stirred up a controversy. One of the main points of contention was Shapley's calculation of distances for the globular clusters. Some astronomers thought that Shapley's measured magnitudes were dimmer than expected because of obscuration by gas and dust between us and the globular clusters, a type of cosmic smog. If so, the distance would be exaggerated, perhaps by a factor of ten or more.

Shapley's distance estimates were also inconsistent, by a large factor, with a less-precise method of distance estimate called *statistical star counts.* In this technique, the relative distances of stars are to be found by postulating an overall shape of the distribution of stars in the sky, and the model is then compared with actual counts of stars at various sky positions to infer distances.

The final point of contention was over the distance to another type of star grouping, called the spiral nebulae. Shapley believed that spiral nebulae were smaller, and closer, than the globular clusters, while others contended that they were separate entities, the island universes. If there were ways to get distances to these spirals, then the issue of the extent, and structure, of the universe might be better understood.

QUEST FOR THE ISLAND UNIVERSES

The spirals are pinwheel-shaped groupings of stars and gas. On the basis of a study made by the Dutch astronomer Adriaan Van Maanen, Shapley believed these spiral nebulae were relatively small and nearby compared with globular clusters. In papers published in 1923 and 1924, Van Maanen claimed that he could see individual stars in some spirals actually making small but perceptible angular shifts, which he believed indicated the motion of the stars rotating around the pinwheel centers. Although this angular shift had a different origin than parallax, parallax had shown that any mea-

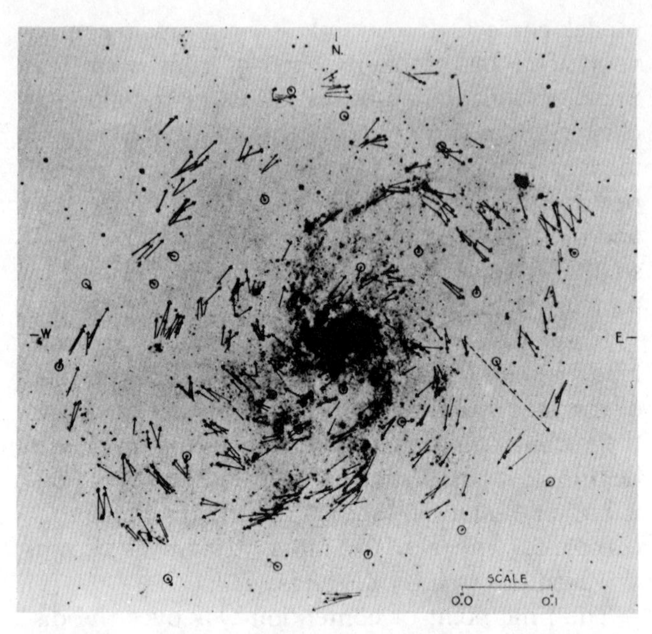

A spiral nebula, with Van Maanen's markings of the rotational motion of individual stars. With a magnified scale, Van Maanen drew lines to indicate the relative swiftness and direction of the motion. The actual motions are only a fraction of those Van Maanen claimed. (Copyright © 1976 Columbia University Press. By permission. Originally appeared in The Astrophysical Journal, *published by the University of Chicago Press; © 1923, Vol. 57, The American Astronomical Society.)*

surable changes of angle must indicate that the object was relatively close by, of the same distance as stars like 61 Cygnus A. Van Maanen's results were strong support for the notion that spirals were far closer than were the globular clusters.

Yet Van Maanen was incorrect. A look at his photographs shows that Van Maanen was claiming the existence of angular shifts that were smaller than photography could record. The problem lay with Van Maanen's interpretation. A typical measurement of a spiral by Van Maanen would indicate a motion for the stars in spirals of twenty-thousandths of an arcsecond per year, and he believed that the uncertainty in this calculation was about one- or two-thousandths of an arcsecond per year. The flaw in his approach was his underestimate—whether by mistake or on purpose—of this uncertainty. In his day, Van Maanen's spiral nebulae rotation studies were considered among the most important in astronomy, and while he held a tight reign on his findings and equipment, they could not be disputed. A few years later, however, the same materials were used to independently repeat Van Maanen's analysis, and the uncertainties were found to be twenty times larger! Van Maanen's studies actually could not be used to make a successful estimate of the distances to spiral nebulae because the uncertainties for rotation were consistent with no perceptible rotation at all.

At about the same time as Van Maanen's angular motion study, another method was being used to set limits to the distance to the spirals.

We have seen that Cepheids are an important tool in calculating distances. But to a less-precise degree, other star types, and other classes of variable stars, share this importance. One of these other types is the nova. *Nova* is Latin for "new," signifying the stars' celestial novelty. These are stars that suddenly appear where previously no star was seen. Slowly they dim again until they become too feeble to detect. A nova's fleeting brightness is caused by the accelerated burning of the star's nuclear fuel, caused by an unstable condition whose nature is still poorly known. Astronomers have some rough clues that novae have a small range of absolute magnitudes at the peak of their burnup, and this means that they offer a crude way of inferring distances, via the distance modulus.

Novae are common in spirals. In 1917 the American astronomer Henry Curtis showed that many of his observed spirals contained novae. He was struck by the profound dimness of these novae, much as Leavitt had been struck by the dim Cepheids in the Magellanic clouds. Curtis found that all the novae in spirals were at least ten magnitudes (over 10,000 times) dimmer than other novae outside the spirals. He used this as evidence that the spirals could not be nearby, although he did not venture actual limits to distance because of the possibility of cosmic smog and uncertainty in the absolute magnitude of these novae at their peak brightness. Curtis felt that the spirals were much farther away than Van Maanen had proposed; indeed, they were far more distant than were the globular clusters.

As defenders of opposing views, Curtis and

Shapley were given a public forum to discuss the size of the universe at a meeting of the American Association for the Advancement of Science in 1920. This forum gave rise to what was later called the Great Debate on the nature of the universe.

Based on his studies of globular clusters, Shapley contended that the universe was finite and consisted of the Milky Way only. All the nebulae were part of the Milky Way and not outside it. The globular clusters bordered the extent of the Milky Way and thus the edge of the universe. Aware of Van Maanen's studies on spirals, Shapley was convinced that the spirals were far closer than the farthest globular clusters. With the opposing view, Curtis maintained that Shapley had found the extent of the Milky Way with the globular clusters, but not the extent of the universe. First, Curtis submitted that the globular clusters were actually many times closer than Shapley thought they were; Curtis cited cosmic smog problems and the emerging evidence that Cepheids come in two types, with slightly different absolute magnitudes at peak brightness. Next he presented the case for the dim novae in spirals, showing that these dim novae had never been seen anywhere else. Curtis argued for a universe whose extent was unknown, one in which the spirals were island universes and the Milky Way was one such island universe.

Who won this debate? Since science is determined by truths rather than sheer eloquence, and the cases on both sides had valid points, the debate was a draw. Its most valuable aspects were not its conclusions (which were ambiguous) but its summary of the observations. Both Curtis and Shapley agreed that a new approach might break the impasse formed by their two points of view.

HUBBLE'S WAY

When he began his study of the globular clusters, Shapley was a young, ambitious astronomer rising through the scientific ranks at a meteoric clip. After years of making observations at the Mount Wilson Observatory in California, Shapley left with a trunk full of photographic plates that assured him of years of fruitful analysis from his new office at Harvard. Additional plates taken in South Africa gave him all he needed to scrutinize the globular clusters from his rotating, pentagonal desk in Cambridge, Massachusetts.

Edwin Hubble was another ambitious astronomer with rural roots, but this is the only similarity he shared with Shapley. Hubble began his career as an attorney and schoolteacher, but later

Edwin Hubble at the 48-inch telescope at Mount Wilson. (Courtesy Caltech Archives.)

returned to his undergraduate love of astronomy. After a Rhodes Scholarship year at Oxford, he went back to the University of Chicago, then worked at Mount Wilson. His was a curious combination of styles, mingling an Oxbridgean manner with a caustic wit—the latter often leveled at Shapley and his model for the universe.

Hubble shared Curtis's interest in the spirals, and he had a hunch that Cepheids might be found in some, even if the spirals were generally quite distant. Ironically, though, it was Shapley's method of distance measurement that Hubble used to find the distance to spirals.

Hubble's study was unique in the quality of photographic plates he took. He was able to discern dim stars well beyond those previously recorded. His talent and patience, along with access to the largest telescope in the world (the reflecting telescope of 100-inch aperture at Mount Wilson), made his prospects for finding Cepheids in spirals favorable. So while Shapley sat and worked, Hubble observed and measured.

Soon his study had uncovered over a dozen Cepheids in his sample of spirals. They were remarkably faint, yet he could still detect their periodic variations.

Using the Cepheid distance modulus technique, Hubble replicated the approach Shapley used on the globular clusters, only this time with the spirals. He described his findings in 1924, inferring distances for the spirals that placed them at least ten times farther away than Shapley's estimate for the farthest globular cluster. With this data, Hubble later usurped the analysis of his Mount Wilson co-astronomer Van Maanen and showed that Van Maanen had made a major mistake in claiming the rotating motion of spirals. Shapley had to concur that the spirals were very distant; spirals were island universes, or galaxies, and the scale of the universe would have to be rethought.

Only a generation after Leavitt's work, investigation of distances with the Cepheid yardstick had shown that the universe was at least a million times more vast than the distance to the nearest stars and that the main components of the universe were galaxies—islands of billions of stars, each containing their own smaller star groupings. Thanks to Hubble, the study of the universe had been redirected, this time toward the exploration of galaxies.

But the Cepheids in spirals pointed out another problem—how to get the distances for galaxies too far to contain detectable Cepheids. What could be learned about the universe at its outermost detectable limits?

REDSHIFT AND HUBBLE'S YARDSTICK

The path to measuring the distance to the galaxies was built on a peculiar property of the spirals—their tendency to have redshifted spectra. This phenomenon was first reported by the American astronomer Vesto Slipher in 1912. By using a prism with his telescope, Slipher broke the light from galaxies into a spectrum of wavelengths. These spectra of galaxies resembled those of the spectra of stars in that they indicated a superposition of "lines" emanating from the hot gases of an aggregate of individual stars. These lines have distinct wavelengths, much as a blue light gives off light at a specific, "blue" wavelength. The wavelength of each line may be a signature, characteristic of the gas present. Viewing various gases in a laboratory with a prism shows the signature lines of hydrogen, helium, and other gases, found to be present in all stars, including the sun. However, the odd property of the spectral lines emanating from galaxies is that they tend to be offset in wavelength from the wavelengths measured in the laboratory. In almost all cases, the wavelengths are longer than those expected; since red corresponds to the longer wavelengths of light, the term *redshifted* was created to indicate this shifting to longer wavelengths.

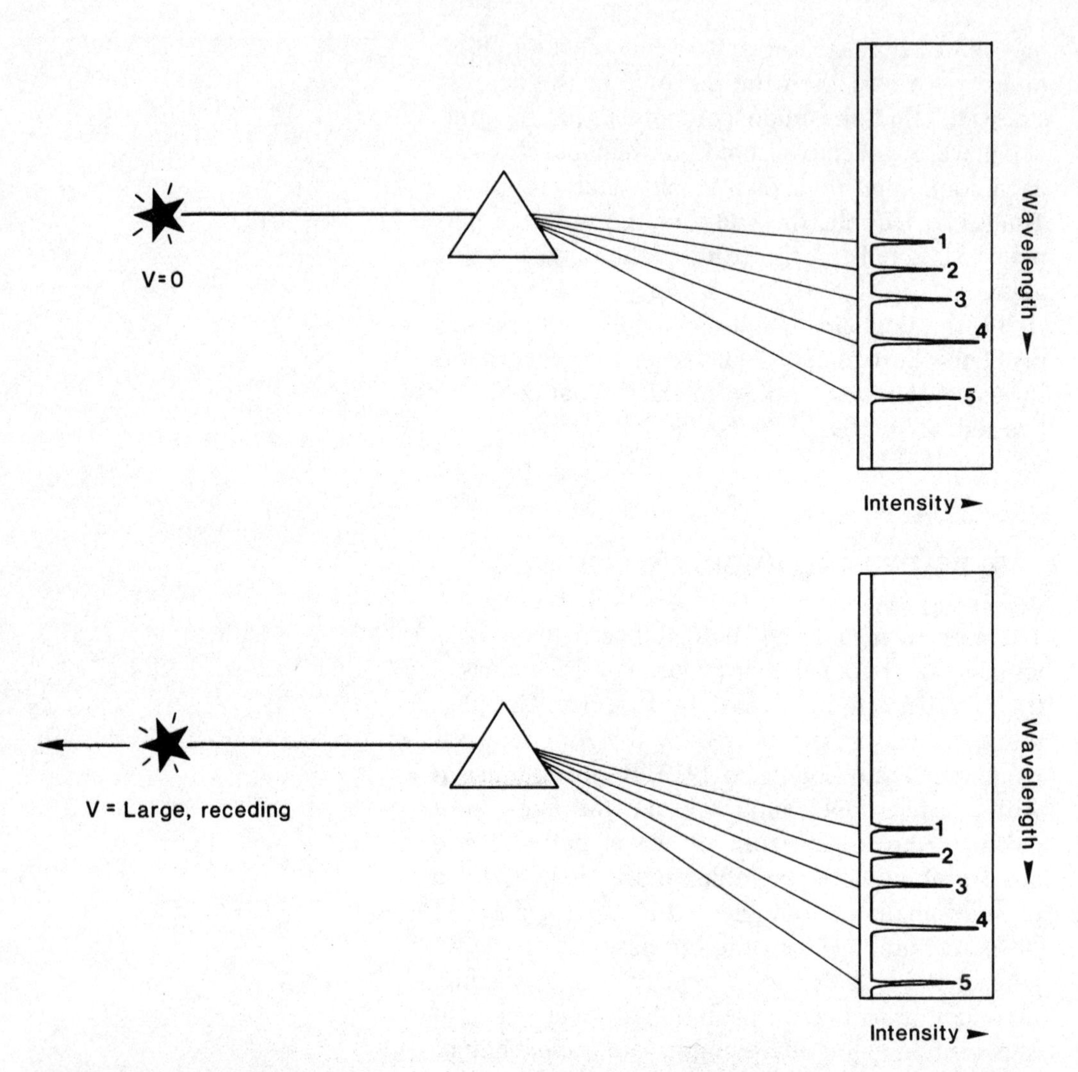

The spectra of two stars, both containing hydrogen gas but moving at different velocities compared with the observer's prism (center). All wavelengths get shifted from the Doppler effect. The amount of the shift can be used to calculate the velocity of the star's motion.

Redshift was, in itself, not an unexpected experience. It is an attribute of the Doppler effect, named after the Austrian physicist Christian Doppler, who discovered it in 1842. The Doppler effect comes about when a radiating object is moving at a velocity either toward or away from the observer. If the speed is toward the observer, the wavelengths get *blueshifted;* if away, then the wavelengths are *redshifted.* Thus a large redshift indicates that the object is moving at a large velocity *away* from the observer.

Although Doppler found the effect first for stars

(he saw stars with red- and blueshifts), it is now known to be a general phenomenon of nature. Consider this example of the Doppler effect: As a fire truck approaches you, the pitch of the siren goes to higher frequencies (shorter wavelengths), followed by a shift to a lower pitch as it passes and moves away from you. Yet the Doppler effect is not a mere curiosity; since the wavelength shifts correspond to speeds and direction, the red- or blueshifts can be used to learn something about the motion of the moving body in relation to the observer.

Redshifted galactic spectra imply that galaxies, almost without exception, are moving away from us at high velocities, most of them at speeds far faster than that of Earth's motion around the sun. Why more galaxies would be moving away from us than toward us was not at all obvious, so Slipher's finding seemed odd at first. But in 1928 the American physicist Howard Robertson plotted the redshifted speed of galaxies versus their distance, focusing on the galaxies for which Hubble, on the basis of Cepheids, had been able to calculate distances. Robertson found that the more distant a galaxy was, the higher redshift it showed, in other words, the amount of the redshifted velocity was proportional to the galaxy's distance. The farthest galaxies were flying away from us at enormous velocities.

Howard P. Robertson, discoverer of the expansion of the universe. (Courtesy Caltech Archives.)

Hubble had not ignored this possibility either, and he came to the same conclusion as Robertson at about the same time. Today, we call this proportionality relation between redshifted velocities and galactic distances *Hubble's law,* and, unfortunately, most have forgotten Robertson's role. Hubble's law is usually written as the equation $V = H_0 D$, where V is the velocity of redshift of a galaxy, D is a galaxy's distance, and H_0 is a constant of proportionality (the slope of the straight line to be found when plotting velocity versus distance) and has units of kilometers per second per million parsecs (megaparsec, or Mpc). H_0 is called Hubble's constant today.

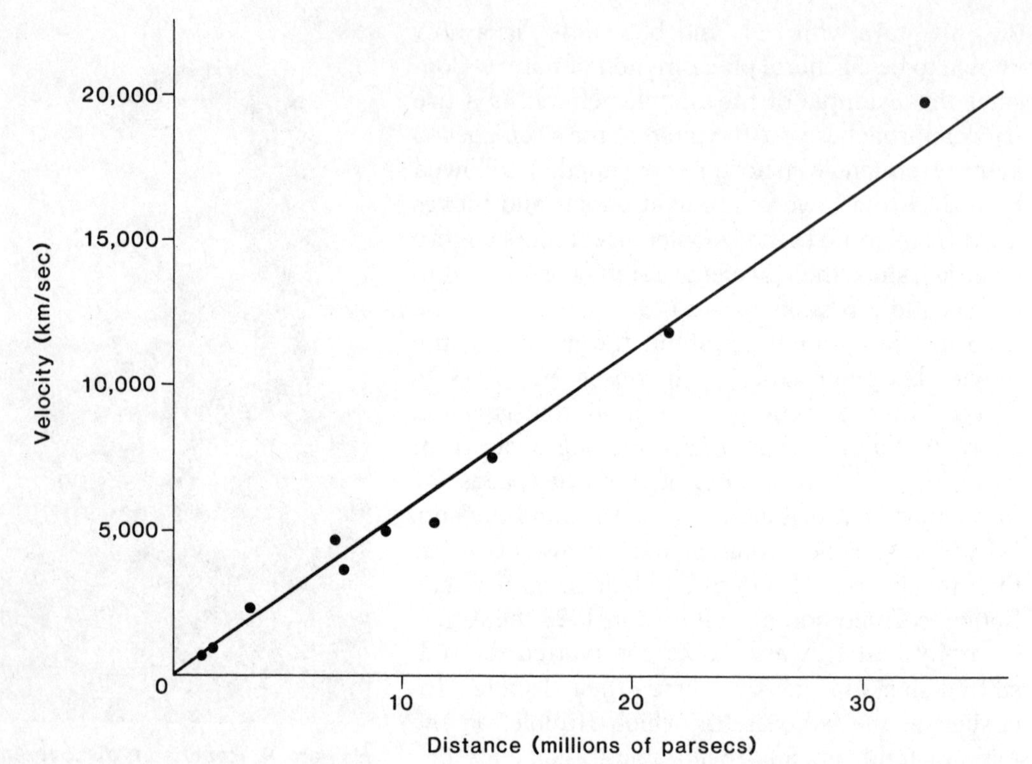

One of Hubble's early versions of Hubble's law. Note the straight-line, or linear, way in which the distances to galaxies relate to the velocity of recession (redshifted velocity). (Courtesy Yale University Press.)

At first glance, the beauty of Hubble's law is that it is an easy yardstick for getting galactic distances. When the value of Hubble's constant is known, then the distance to any galaxy can be found by taking a spectrum and determining the redshifted velocity. This means that galaxies far beyond those from which Cepheids could be detected were within the realm of a distance estimate. And, based on the spectra of many faint spiral galaxies, Hubble found distances dozens of times greater than those in his Cepheid study, without even touching on the countless galaxies that were too dim for him to measure their spectra. The universe was constructed on a grand scale indeed!

Yet to consider Hubble's law merely a tool for distance measurement is to ignore something profound about the universe itself. What is it about the universe that makes galaxies fly away from one another?

BIRTH OF MODERN COSMOLOGY

People have speculated about the origin and evolution of the universe since long before recorded history. But it takes observations to sort through conjectures so that only realistic ideas remain. And after the discovery of Hubble's law, only theories that accounted for the fly-away universe could have any credence.

Naive views of the universe—that it is infinite and eternal—do not take Hubble's law into account. Rather, they assume that the universe is ordered and relatively static, or unmoving. Just as Shapley's model for the universe was overturned by the discovery of the nature of spirals, the concept of an unmoving universe was overturned by Hubble's law. Olbers' paradox, in itself does not address a moving universe and so is of limited use in finding out something fundamental about the universe.

The German mathematician and physicist Albert Einstein showed in 1915 that his theory on gravity—called *general relativity*—held great promise in helping to understand the workings of the cosmos. General relativity contends that the fabric of the universe—the dimensions of space and time—are intimately entwined with matter and its gravitational force. And since gravity is truly the only universal force, where each bit of matter is coupled to all others over all distances, gravity rules. Einstein treated gravity within a complex framework of mathematics and solved the problem of understanding gravity through a series of "field equations." The value of Einstein's field equations is that they make a strong prediction about the universe: It is never static; it can only be in a state of contraction or expansion. This notion hardly approaches an intuitive or common sense notion of the universe; only the delicacy of Einstein's insight and mathematics made this idea tenable.

Einstein's ideas were the prime example of

the theory side of cosmology, where a basic understanding of the observations is coupled with physics to make models, or math-driven explanations, to account for what we see and what we should look for to get a more complete understanding. Yet Einstein's field equations also illustrate one of the issues of this modeling approach. The math may indicate more than one explanation (just as some algebra equations have multiple solutions), while the universe corresponds to only one of those possible solutions, if any at all. What emerges from genius, pen, and paper has to jibe with what we actually observe.

To a certain extent, Einstein's field equations were too general. It remained for other theorists to pick out particular solutions to the field equations and to see what they predicted for the universe's course. Although several physicists made progress on this front, it was the Belgian as-

George LeMaitre (center) with Albert Einstein (right), visiting Caltech in the 1930s. Caltech president Robert Millikan is at the left. (Courtesy Caltech Archives.)

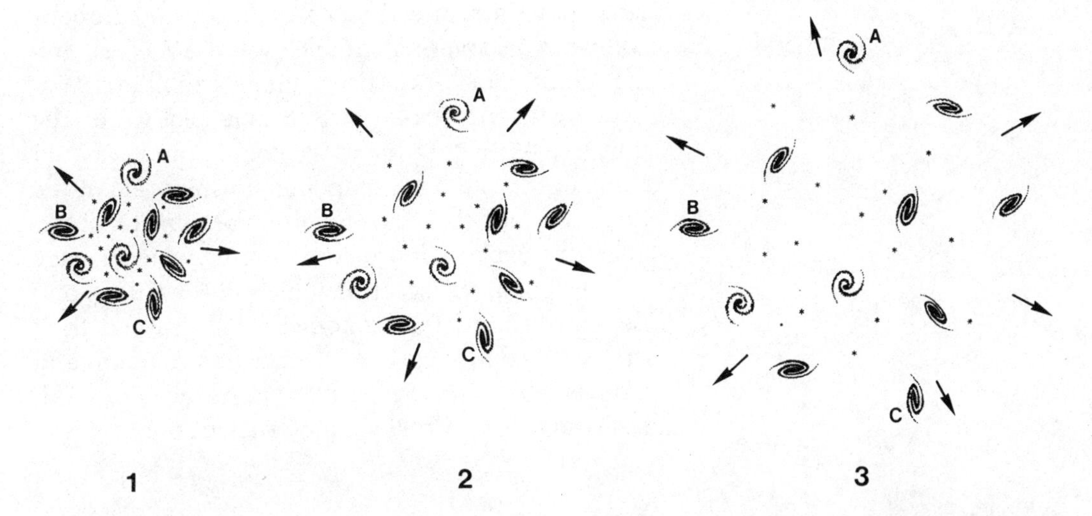

LeMaitre's expanding primeval atom universe at three points in time, showing how galaxies A, B, and C move apart.

tronomer George LeMaitre who came up with an elegant model that explained Hubble's law and placed the observations firmly in place with a theoretical basis for understanding the universe.

George LeMaitre was an astronomer, civil engineer, and Jesuit priest—a man of diverse abilities and great scholarship. After studying Einstein's theory of general relativity, LeMaitre constructed a model for the universe that had an uncanny resemblance to aspects of the biblical story of creation. He contended that the universe had a beginning, and this beginning was as a gigantic "primeval atom" that exploded, causing its pieces to be flung in all directions. According to LeMaitre, Hubble's law occurs because we are observing the expanding universe *within* the expanding universe.

To use a simple analogy, consider the distance between many different points on the skin of an expanding balloon. As the balloon gets bigger, the distance between the points increases, and the greatest increases occur between points farthest away from one another. With galaxies, the expansion of the universe causes the farthest galaxies to be moving away from us at the greatest rate. Hubble's law is therefore an observable consequence of an expanding universe.

LeMaitre's primeval atom model for the universe was introduced just as Hubble and Robertson were discovering Hubble's law; LeMaitre's timing could not have been more propitious. For here, with one stroke, was a bold picture of the universe—one that had a beginning, and was not static or infinite but expanding. Einstein said of the LeMaitre model, "This is the most beautiful and satisfactory explanation of creation to which I have ever listened." Between Hubble's observations and LeMaitre's model, astronomers had grasped their first view of the secret of the universe's origin and nature—the perfect high point to decades of modern progress and millennia of ancient ones.

· TWO ·

After Hubble—Quasars, CBR, and the Standard Cosmological Model

IMAGINE Edwin Hubble's great fortune. He had proved something profound and fundamental about the universe—that it was expanding—and in the process provided a new way to infer the distances of the millions of distant galaxies. Part of it was good luck: He had access to the world's largest telescope and was able to study the galaxies with unprecedented exactitude. It is easy to understand the confidence, perhaps hubris, that drove Hubble and his colleagues from that point onward. Careful studies of galaxies could yield even greater insights into the universe. But new discoveries were also to complicate the simple approach of Hubble's law, as new secrets of the expanding universe were revealed.

To expand his observations, Hubble needed to find another distance yardstick, like the Cepheids, to see how very distant galaxies held up to the Hubble law. But beyond the Cepheid yardstick lay a no-man's-land of distance-determination methods, all less precise, to varying degrees, than the Cepheid method. Hubble progressed with an assumption: Galaxies with similar shapes (spirals, for instance) might have the same absolute magnitude. Then, through the use of the distance modulus, their measured magnitudes should provide distances. In effect, Hubble was able to equate the distance of a galaxy with its dimness. As a convenience, he replotted the diagram of redshifted velocity to distance with redshifted velocity to measured magnitude, a figure known today as the Hubble diagram. Within this framework, Hubble proceeded to see what else the receding galaxies could tell us about the expanding universe.

The Hooker Telescope at Mount Wilson, California, used by Edwin Hubble to explore the expanding universe. (Courtesy Caltech/Mt. Wilson Observatory.)

When understanding the distant galaxies, Hubble and his colleagues had to take into account a subtle point that is quite removed from any worldly experience—cosmic time travel. To look at the farthest galaxies is to look back billions of years in time. This happens because light does not reach us instantaneously; the light from the farthest galaxies takes many billions of years to reach us. We are seeing views of these farthest galaxies in their childhood and adolescence, whereas the nearby galaxies show us older views of themselves because their light takes only millions of years to reach us. Yet we see the farthest galaxies at points in the universe's growth when its expansion was much younger. Looking out to great distances means looking back to a time when the universe could have actually been expanding faster than it is now. How this should affect the appearance of the Hubble diagram was as follows: The closer galaxies (bright measured magnitudes) should line up along the

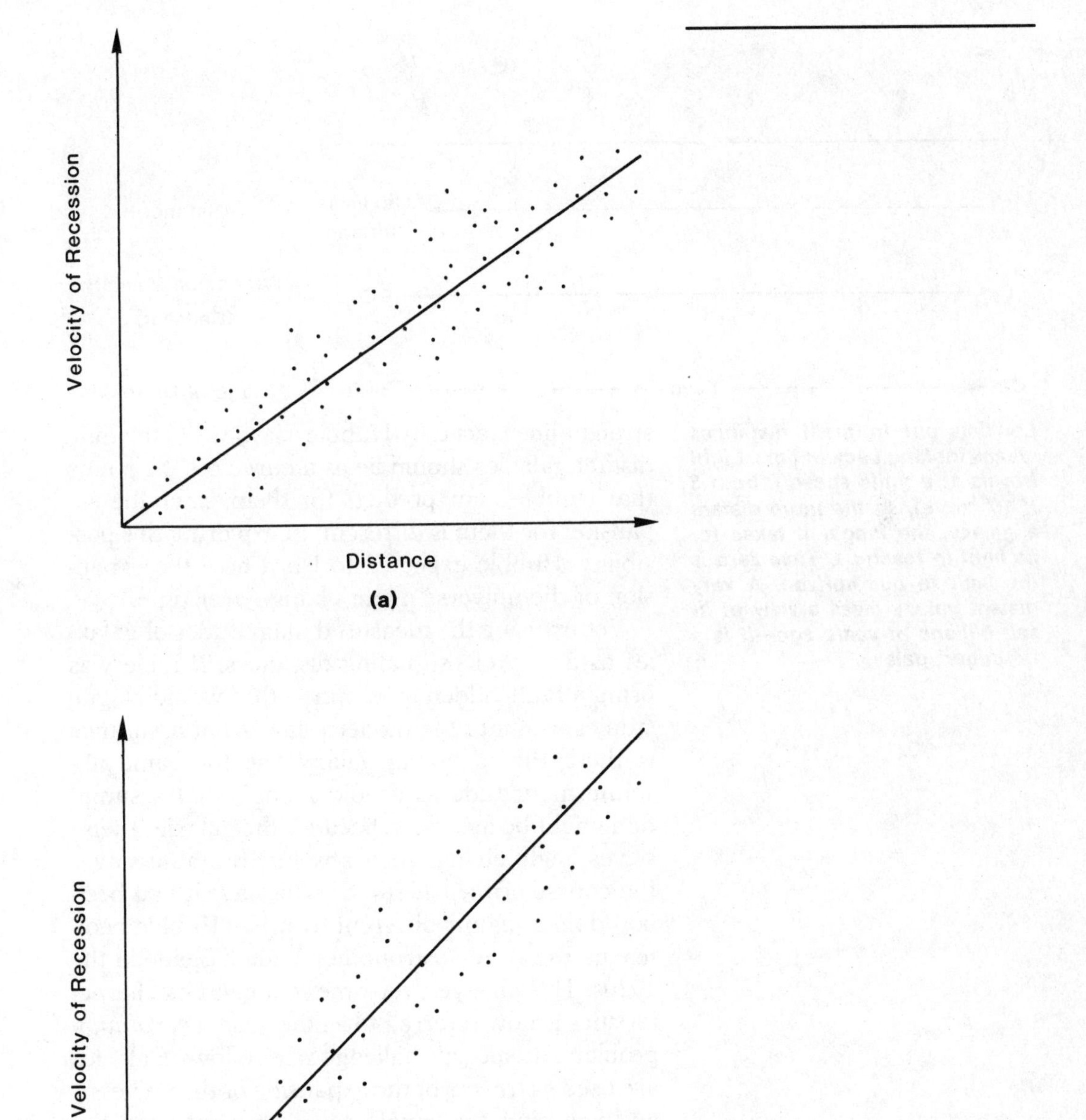

The Hubble diagram in two forms: (a) velocity of recession and distance; (b) velocity of recession and measured magnitude.

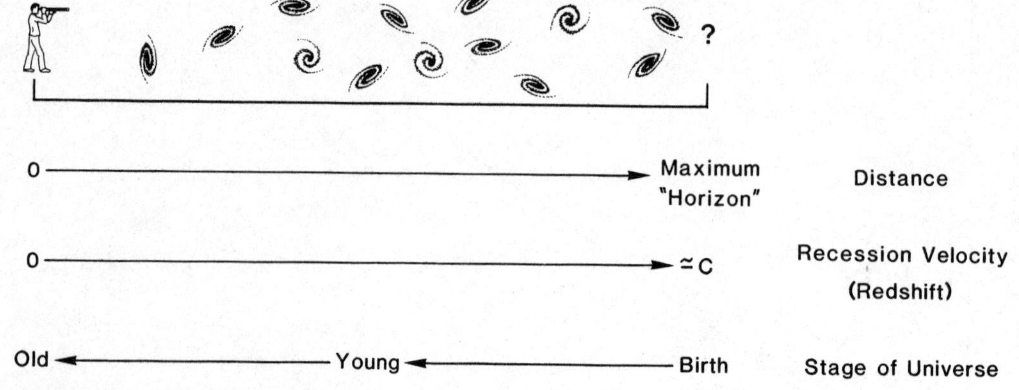

Looking out to great distances means looking back in time. Light travels at a finite speed (about 3 $\times 10^5$ km/s), so the more distant a galaxy, the longer it takes for its light to reach us. Time zero is the limit to our horizon. A very distant galaxy gives a view of itself billions of years ago—it is a "younger" galaxy.

straight line traced by Hubble's law, while the dim, distant galaxies should lie as a curve off the points that Hubble's law predicts for them, since the expansion for them is different. In exploring this possibility, Hubble expected to learn how the expansion of the universe might change over time.

Yet by using the measured magnitudes of galaxies as a way of estimating distances, Hubble was using a fault-ridden assumption that would plague studies well into the modern day. What assurance is there that a young galaxy has the same absolute magnitude as an older one? That assumption might be incorrect, because the galaxies themselves might change their absolute brightness over the course of their lifetimes. This caveat had been posed (in a slightly different form) by Hubble's colleague the Dutch astronomer Walter Baade in the 1930s. This change over time in a galaxy's characteristics is now referred to as the *galactic evolution* problem. Its special challenge is as follows: Galaxies are used as tracers of the expansion of the universe, an expansion that might have changed over the course of billions of years. But the galaxies themselves might also have changed. If the measured magnitudes of galaxies are to be used to trace the course of the universe, how can one separate the magnitude of an individual galaxy at a particular evolution stage from the effect one might expect from the universe's expansion? Does the study of the farthest galaxies via magnitude really offer pos-

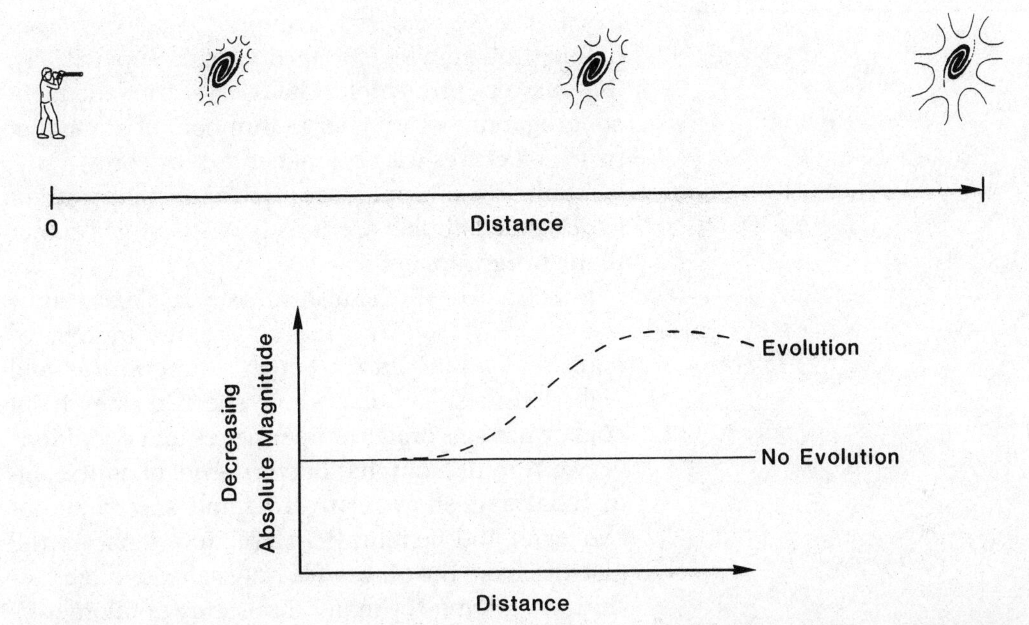

Galaxy evolution—the change in a galaxy's properties with time, independent of how the universe expands. Are younger galaxies brighter or dimmer, as shown by the dotted line?

sibilities for finding out more about the evolution of the universe?

Hubble went ahead with his work, despite the limitations of assumptions about the lack of change of galaxies over time. This was not altogether unjustified; he thought that only the most extremely distant galaxies could possibly be affected by galactic evolution. And Hubble's extension of his early results enforced the validity of Hubble's law. But as dimmer galaxies were studied, they were found not to lie on the straight line of Hubble's law, or on any other curve. Instead, they scattered. When Hubble's pioneering effort ended in the 1950s, his students continued his work. But by the early 1960s it became clear that the younger universe was one where galaxies might be different. That difference was to be exemplified by an entirely different category of distant object—the quasars.

THE COMPLICATED QUASARS

How should we define a galaxy? Is it just a highly structured group of millions of stars? What features can we see that distinguish galaxies from nebulae or smaller star groupings? Superficially, the answers are simple. Galaxies are independent congregations of very large numbers of stars, that form structures with bright centers, or cores. Stars, in comparison, appear as individual pinpoints of light, while nebulae are mostly wisps of gas rather than star groupings.

Spectra, too, distinguish galaxies. Galaxies show spectra that look like the aggregate starlight of many individual stars, at high temperatures and high densities. Nebulae show spectral lines from cold or hot gas, under temperatures and conditions very different from that of stars. And, of course, individual stars show relatively simple spectra of hot hydrogen and helium. Redshifts, too, separate the galaxies from the others, for only galaxies have redshifted velocities of many thousands of kilometers per second, or more.

So consider the following puzzle. An astronomer takes a photograph of a patch of sky, compares it to an earlier photograph, and notices that a star has changed its magnitude. Suspecting a variable star, the astronomer proceeds with a spectral study. But rather than finding a typical stellar spectra, he finds what resembles a very hot nebula with large internal motions. But he knows it is not a nebula because the redshifted velocity of the spectra is huge—what one would expect from a very distant galaxy. The object looks like a star and shows the spectral lines of an exotic nebula, but it is receding from us as a galaxy does. This odd object is a quasi-stellar object, or *quasar.* The first quasar was identified by Caltech astronomer Maarten Schmidt in 1963. Others soon followed, and by the mid-1960s many dozen were known. Each quasar showed the huge redshifts expected of distant galaxies, and it is this characteristic that placed them as objects outside our galaxy. But quasars are not galaxies in the normal sense—their extraordinary behavior is totally unlike that of galaxies.

In astronomy, extremes often prove the most interesting, and it is easy to use superlatives in describing quasars. They are, for example, the most intrinsically bright objects in the universe. The light one quasar gives off is equal to the light given off by dozens, sometimes hundreds, of ordinary galaxies. It is as if many hundreds of novae erupted sequentially and simultaneously to light each quasar. Yet all this violent activity occurs over a very small region of space.

We can use a simple exercise to determine the size of a quasar, based on its variable magnitude over time. A change in magnitude that occurs over a few days implies that any energetic disturbance (which causes the change) can travel only a few light-days at most. Based on this variation, a maximum limit of 100 or so times the size of the solar system is a typical estimate of a quasar's size. An incredibly bright, violent, and compact environment spells out the nature of quasars. And since the spectra have nebula-like characteristics, these quasars are not just very dense star regions but ones where huge regions of gas are ionized, superheated, and tossed about. Quasars are a hell for stars—or what remains of them after their disintegration.

Spectacular as a class of objects, quasars are a fundamental puzzle in cosmology. For one thing, Hubble's diagram, and its assumption about the absolute magnitude of galaxies, doesn't hold up for such bright yet varying objects. How could one accurately (say, to a factor of four or so) use their magnitudes to get their distance? The infernal conditions within quasars are radically different from the sedate cores of galaxies. And because of the violent changes taking place inside them, quasars may vary in absolute magnitude over a large range.

Quasars epitomized the problems astronomers encountered in trying to probe the greatest distances in the universe. For to look out to the greatest distances, the dimmest galaxies need to be studied. But at such great redshifts, the comparatively bright quasars are the main objects to be seen.

Rather than being an easy extension of Hubble's first studies, the problem becomes one of understanding the distant galaxies and quasars in order to also comprehend their role as probes of the early universe. Hence, extending Hubble's law to the reaches of the observable universe might not be the best way to probe the cosmic expansion.

Yet quasars and galaxies might not be the only probe of the expanding universe. The next hint of its workings was to come from an entirely different perspective.

CBR—THE REMNANT FIRE OF CREATION

Today there are thousands of cable TV antennae (dishes) in common everyday use. These dishes are capable of picking up much more than game shows or videos, however; with the appropriate extra equipment, one can perform an observation revealing a profound piece of information about the history of the universe.

And it all comes out of the noise.

The observation entails nothing more than measuring the amount of noise coming from the sky, which the antenna picks up when it looks at different randomly chosen sky patches. At certain frequencies, the entire sky is a source of faint but diffuse radio noise; it is a weak but all-pervasive radio emitter. The sky noise isn't much to look at, since only an occasional bit of TV "snow" comes from the sky itself. Yet while this noise is highly limited in entertainment value (save for those who are easily amused), it nonetheless is intriguing from a scientific standpoint.

Oddly, the feeble intensity of the noise will be virtually constant, regardless of the direction at which you point your dish. This aspect—the lack of directionality of the radio noise—is but one of many. Another is the noise's tendency to remain constant no matter what the time of day, year, or weather condition. And the amount of noise varies

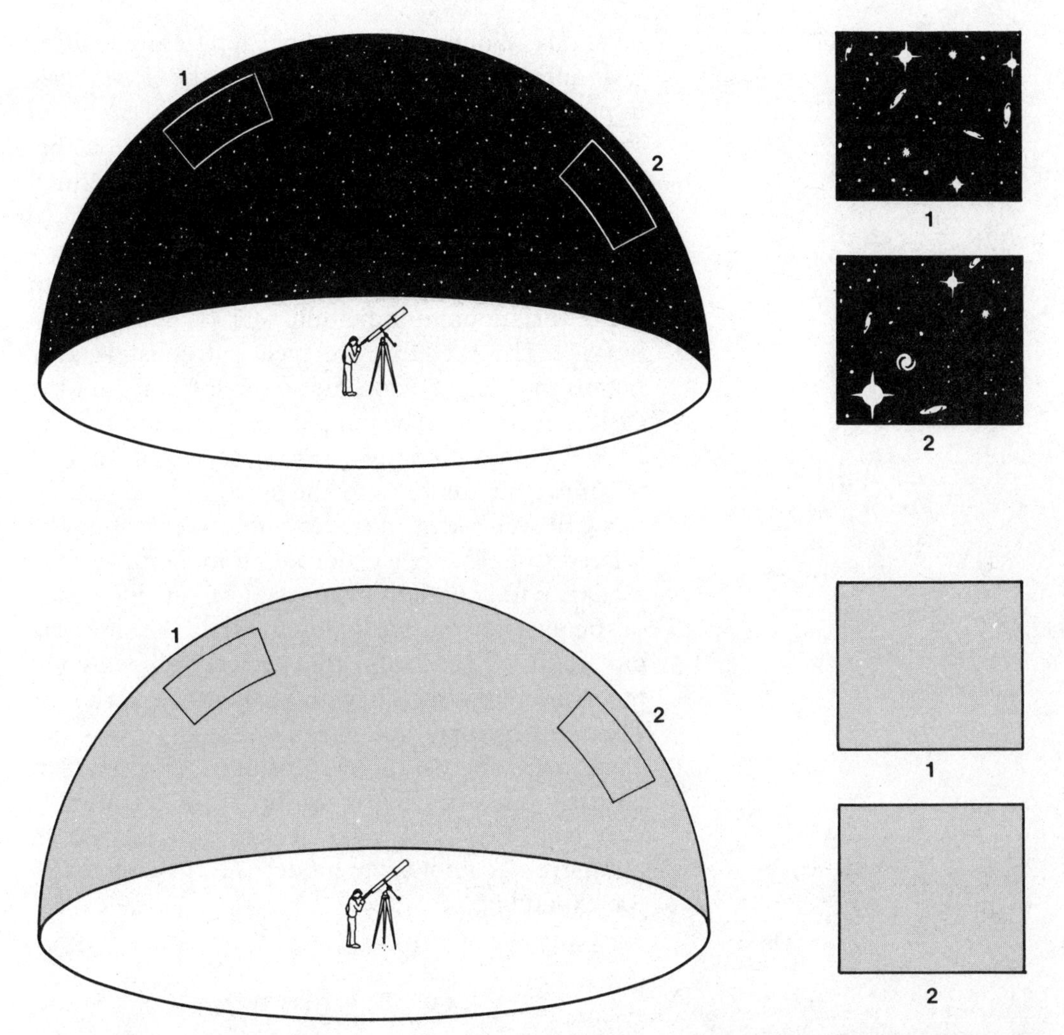

in intensity with wavelength, as if it were exhibiting a small range of radio "color." Because this noise is so constant, it is referred to as a background. And because its constancy is unaffected by Earth and its atmosphere, it must be from far beyond Earth; a cosmic background.

An observer (top) looking at the sky at visible wavelengths sees that two patches of sky look very different. But an observer with "microwave eyes" sees a nearly constant, diffuse glow, virtually identical at the two patches.

This *cosmic background radiation* (CBR) was first discovered in 1964 by American engineer-scientists Arno Penzias and Robert Wilson while they were exploring the communications link to the Telstar satellite. Penzias and Wilson weren't searching for cosmic meaning or TV reruns; their goal was to enhance the efficiency of satellite communications. What set apart their apparently routine task, and what ultimately won them the Nobel Prize in physics, was their determination and success in showing that the noise was celestial, as well as being the same in all sky directions, which is defined as *isotropy.*

The visible universe is one that shows clumpy and random orderings of stars and galaxies, with great variations in individual star or galaxy magnitudes. The fact that we pick out constellations means that the sky does not *look* isotropic. Yet the CBR *is* isotropic. The simplest explanation for the CBR was that it comes from a very early stage of the universe, far beyond the stars and galaxies. If this CBR represents an infant universe, it could be related to LeMaitre's primeval atom, before it had a chance to fragment into galaxies. Yet since each of the parts of this early universe has the same radio "color," we can infer that they are all heated to the same temperature, with virtually no variation. This temperature, though, is exceedingly low—very close to 3 degrees above absolute zero. Here is an early universe that is uniform and, apparently, very cold. Does this make sense in the context of LeMaitre's cosmological model and the concept of the expanding universe?

GAMOW'S BIG BANG

In 1948, long before Penzias and Wilson's discovery of the CBR, an important paper had been published hypothesizing the CBR's strange existence. The prediction, however, sunk into obscurity, despite the excellent reputation of the physicist and collaborators who devised it. The physicist was George Gamow, a Russian-American, and the details of the theory's model later came to be called the Big Bang.

Gamow devised a cosmological model that was a follow-up to LeMaitre's earlier work. Gamow wished to understand what the very early universe, when the expansion was still in its infancy, was like.

At this time, the universe must have been so compressed that incredibly high temperatures and densities prevailed. Curiously, a field that had little to do with cosmology struck Gamow as the perfect arena for understanding this early universe.

That field was nuclear physics, fresh from the famous (and infamous) success of the atomic bomb. Gamow showed that what had been learned about atomic nuclei from the atom bomb and other studies could be used to show what conditions were like in a much bigger, and profound, "explosion"—the beginning of the universe. In particular, Gamow was able to show that hydrogen and helium were made under very hot temperatures and very high densities, such as those encountered in an atomic fireball or the beginning of the universe.

Judging by its unimaginable compactness, Gamow believed the universe's start must have

Robert Wilson (left) and Arno Penzias (right) in front of the microwave horn (an antenna much like a dish) used to discover the CBR. (Courtesy Bell Laboratories, AT&T Archives.)

been a time when the density and temperatures were extreme. This very young universe was really a universe of radiation and young elementary particles—fundamental bits of matter—a soup in which matter and radiation were the same. As the Big Bang occurred, its increasing volume led to a decrease in temperature and density. It is these changes in density and temperature that led to a series of ever-changing states for the radiation-matter soup, changes that eventually led to the formation of hydrogen atoms and slighter amounts of helium. But one stage of this changing soup was especially interesting from the point of view of what we, billions of years later, could observe.

Sometime well into the expansion of the primeval atom, the temperature had decreased enough, along with the density, that the soup, once opaque to light, became thin enough to let light escape. This is the very last stage in which radiation is said to "dominate" the young explosion.

What started at greater than many billions of degrees in temperature has cooled at this stage so that the emerging hydrogen is only 4,000 to 6,000 degrees, a point at which most of the radiation should appear as visible light. Gamow believed this point was a key signature of a version of LeMaitre's theory of the primeval atom, where a very hot, infant universe expands and cools. This moment in time, however, is not just an unwitnessed piece of

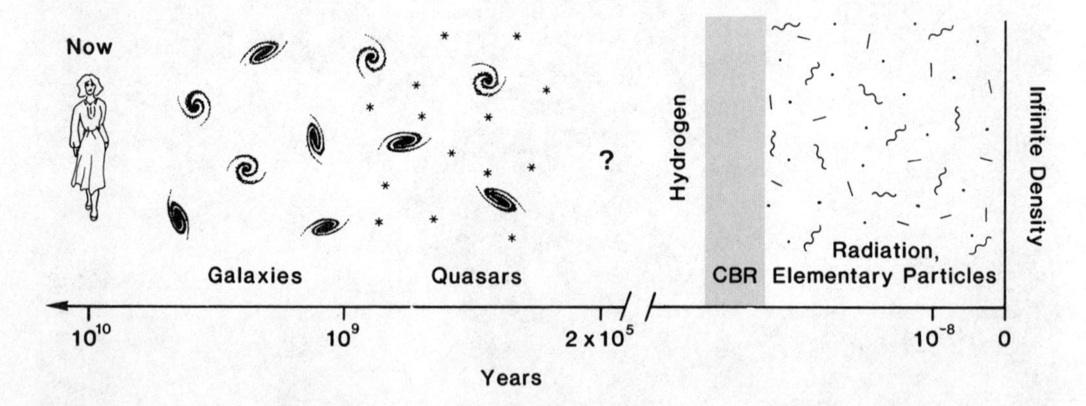

A simple view of the growth of the universe, as suggested by Gamow's model.

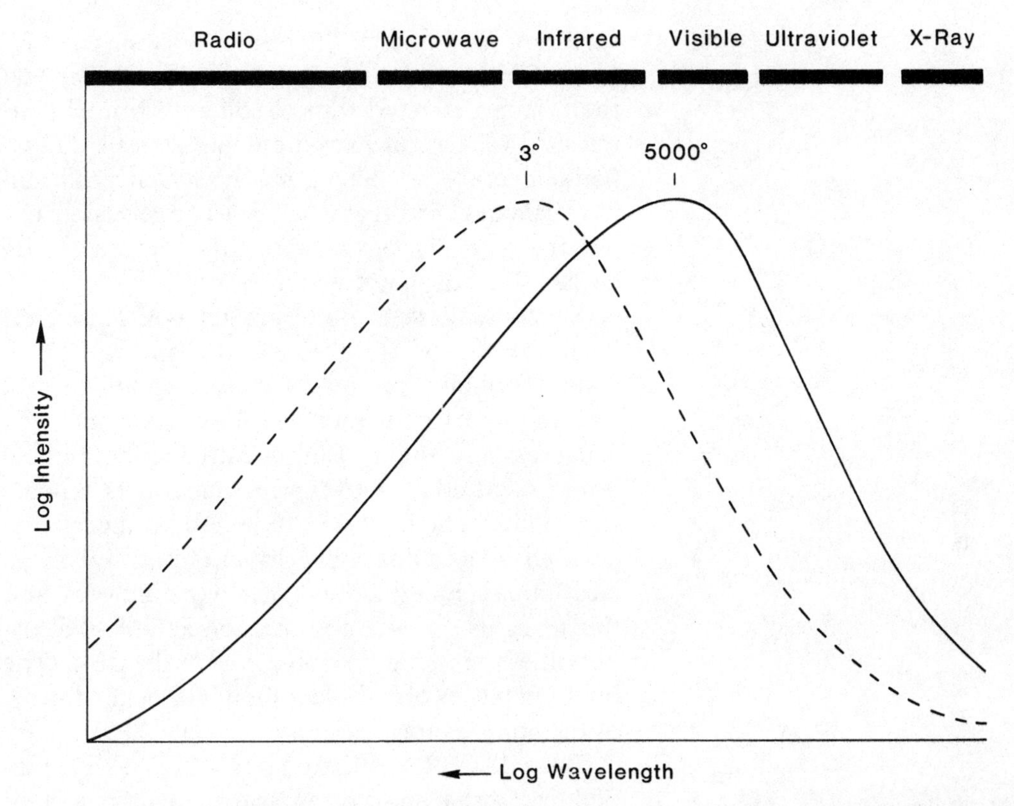

cosmic history. By looking out to the many billions of years ago when this stage occurred, we can see this early universe stage, an omnipresent haze of radiation, everywhere the same. This is the CBR witnessed by Penzias and Wilson.

Even though the CBR emits most of its radiation at light wavelengths, the huge Doppler shift of the CBR relative to us makes it appear in the microwave and infrared wavelengths, which correspond to the regimes of far cooler emitters.

Two curious puzzles about the CBR, as a remnant of the fires of the Big Bang, become apparent with a little thought. First, the 3 degree temperature of the CBR is hardly the 4,000 to 6,000 degrees of the shell. Yet with the great time difference between our stage of the Big Bang and this earlier one, the CBR should be at a great distance—and thus at huge redshifted velocity! Because of this redshift all the wavelengths of the shell's radiation show up at much longer wavelengths. What we measure as a source of radio radiation from a 3 degree temperature is actually the very highly redshifted visible-light emission of this point when radiation had its final fling in dominating the universe. So the apparent frigidity of the CBR is just the Doppler-effected shift of something much hotter. The CBR is an extreme expression of the expanding universe, before the explosion broke up into galaxies. Far from being a phenomenon unlike the recession of galaxies, CBR represented an earlier stage of that phenomenon.

A second puzzle is the CBR's isotropy. This might be illustrated by the bubble gum analogy. Here, a small bubble, continually increasing in size, will have any part of its surface appear virtually the same as any other. But eventually the bubble bursts, scattering sticky pieces about its embarrassed host. The Big Bang, although far more complicated, shares this similarity in uniformity. At its early stages, its exploding parts were uniform, and this gives us the isotropy—the directionless distribution of the radiation—observed in the CBR. (The bursting pieces of bubble gum might be analogous to the galaxies formed at a later stage.)

This odd CBR, postulated by Gamow, and identified by Penzias and Wilson, was thus an important clue about an earlier stage of the expanding universe. Unlike Hubble's law, however, the CBR tells little about how the expansion evolved; as one point in time, it was not a good indicator of how fast the expansion occurred.

BIG BANG—PARAMETERS FOR THE STANDARD MODEL

Not all that astronomers want to know about the universe is locked up in the CBR and Hubble's law. It would certainly be interesting to know what started the expansion and what factors have controlled it over the billions of years since the Big Bang started. Here we shall see that the Big Bang, as a model, has two main variations; deciding which of these is most pertinent to our universe

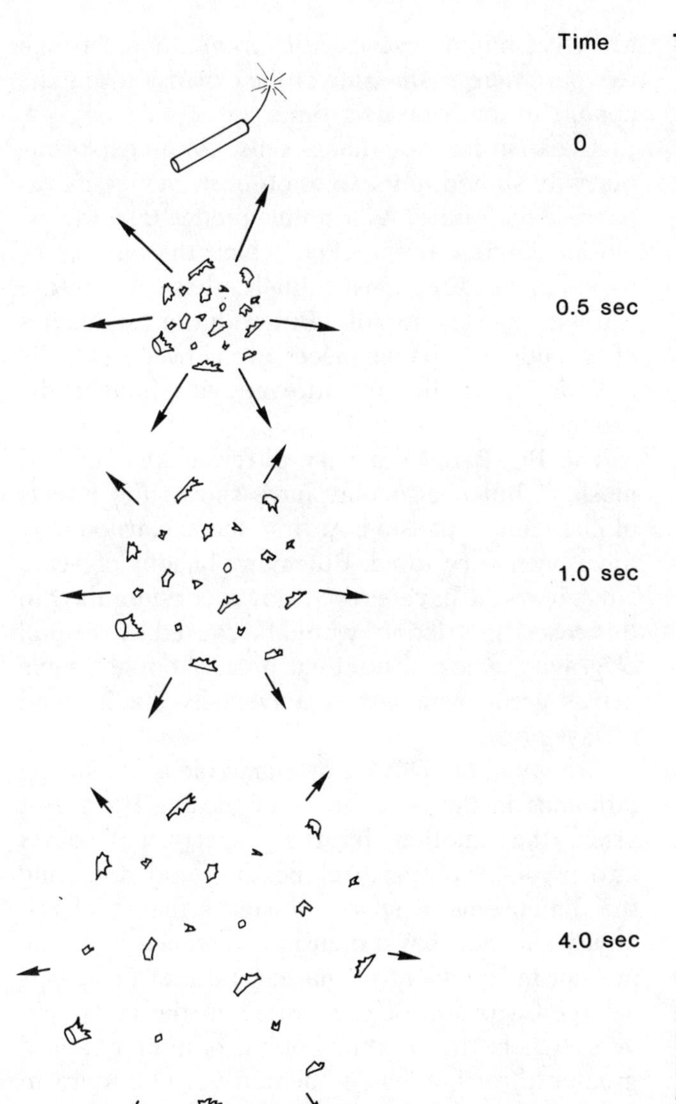

The exploding firecracker starts with a great velocity of expansion, but braking by the air slows it, just as gravity may slow the expansion of the universe.

is a major goal of the observing cosmologist. These variations come about as possibilities of LeMaitre's primeval atom and Gamow's follow-up, and they represent an explanation that has become standard during the last twenty years and that encourages observers to seek out its predictions.

Variations on the theme of a Big Bang differ mostly on how the expansion has progressed over time. We might describe this progression through two parameters—the rate of the expansion, and the change in that rate over time.

An expansion rate makes sense for an explosion, but why should any explosion ever change its expansion with time? As a simple model, take the explosion from a firecracker, which throws bits of paper in all directions. Initially, these pieces are thrown out very rapidly. But within a few inches of the pop, the paper pieces are slowed down by the friction of the air and eventually float to the ground.

The Big Bang is a very different kind of "explosion," but it also may have shown the effects of changing expansion. At first, the expansion may have been very rapid. But many billions of years later, it could have slowed down considerably. In this case, the "friction" would be caused by the pull of gravity, whose attracting force, although feeble across great distances, is universally binding and always present.

Gravity, then, plays a primary role as a slowing influence in the later stages of the Big Bang. But gravity has another, bizarre property that comes into play—its property of bending and distorting the dimensions of space. Einstein's theory of relativity showed that dimensions of space and time are not independent of matter and that there was a large distortion of dimensions in the early universe, where the proximity of bits of matter caused greater distortion than in the universe of our era, in which objects are farther apart. This gives the universe its *curvature.* Curvature is slight, but it can make very distant galaxies look different from local ones in size and shape. Matter—the universe as a whole—influences space, and the degree of that influence depends on how much matter there is. These two effects of gravity determine how the expansion of the Big Bang looks to us.

The two expansion parameters are commonly called the expansion rate and the deceleration parameter. The *expansion rate* is the change in

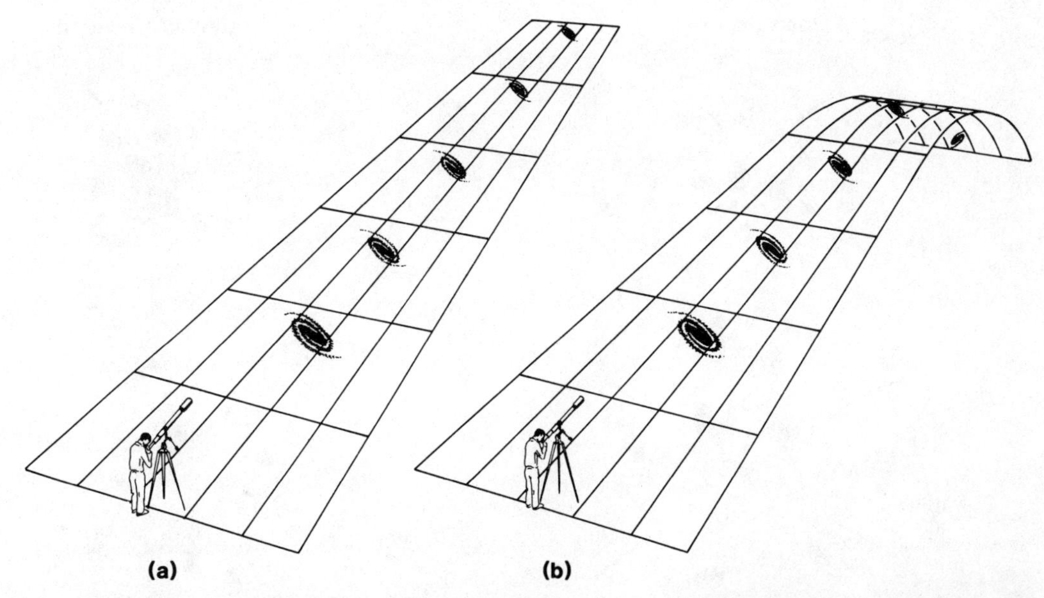

In a Euclidean, or flat, universe (a), dimensions of space act as we expect them, while in a curved universe (b), gravity distorts the dimensions over great distances.

the velocity of expansion with increasing distance. It is actually the same as Hubble's constant, H_0, and the two are sometimes discussed interchangeably. The expansion rate is likely to be different now than it was many billions of years ago, and this gives the potential for deviation from the straight line of Hubble's law. This represents the braking, or *deceleration parameter,* caused by the effect of gravity slowing the expanding universe's rate. The deceleration parameter is symbolized by q_0.

Since the amount of matter in the universe governs its expansion properties, a simple model equation describes this amount of matter and compares it to the expansion rate and deceleration parameter. It is called the Friedmann equation and is given as

$$H_0^2 = \frac{4\pi G}{3} \frac{\rho}{q_0}$$

where π is a constant, as is G, and ρ, the Greek symbol rho, is the density of the universe—the amount of matter per unit volume. In concrete terms, this equation relates how the density of the universe

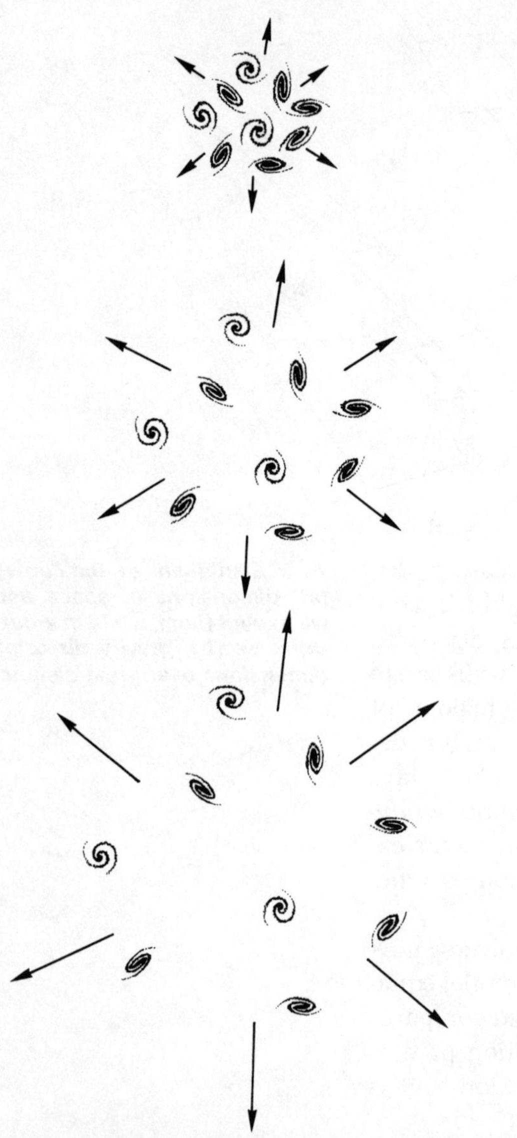

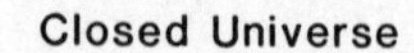

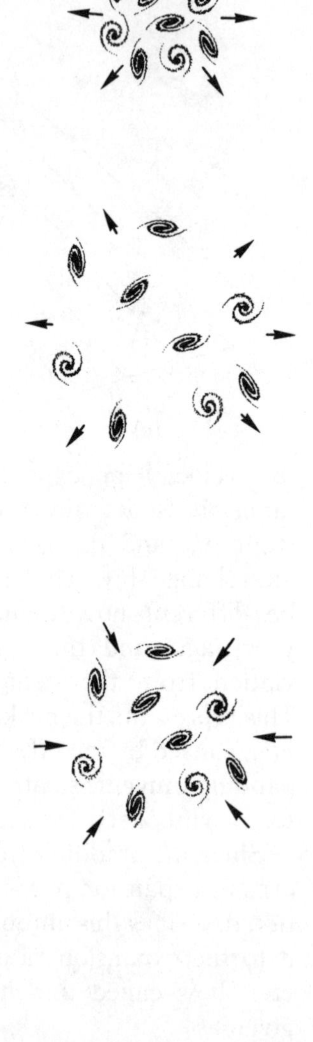

An open universe (left) expands forever, while a closed one (right) may stop and contract.

and its expansion rate are controlled by the curvature, through braking, expressed by the deceleration parameter, q_0.

It is the possible values of H_0, q_0, and the density that give the Big Bang variations. Basically, the distinction between these variations is how the universe is expanding: Is it expanding so fast that it goes on like that forever, or is gravity strong enough to pull the pieces back together—an implosion—at a later time? Or is the universe a delicate balance between the two so that it expands, slows, and contracts, only to expand and recycle again?

Right now, the universe seems to be only expanding, so there is little doubt that the expansion shall continue for some time to come. But if the expansion was faster earlier on, we have strong evidence that it will change over time. By looking into the past, we can predict the universe's future course.

One useful concept in this expansion/contraction idea is that of the universe being open or closed. In an *open universe,* there is not enough matter to prevent the expansion from going on forever, while in a *closed* one, there is sufficient matter to actually counteract the expansion, through gravity, causing the universe to ultimately contract. Whether the universe is open or closed depends purely on how much matter it contains. Along with this concept of openness or closeness, there is the notion of a *critical density,* ρ_{CRIT}, a parameter that registers how much matter the universe must have for it to stop expanding; critical density is this total amount of matter divided by the volume of the universe. There are specific values of ρ and q_0 that are determined by the opennness or closeness of the universe, so that finding these values may solve the issue of which Big Bang—open or closed—we live in. And as we see from the Friedmann equation, Hubble's constant must be consistent with the values of q_0 and ρ for these values to be correct. In a very real sense, cosmology today is a hunt for these magic numbers.

One final cosmological parameter relates the time scale for the expansion—how long the universe has been expanding. Because the braking effect of deceleration can slow down the expansion, the time scale for the expansion of a closed universe differs from that of an open one. We speak of a Hubble time, which is defined as the inverse

Table 2.1 The cosmological parameters for an open and closed universe according to the Big Bang model

	Open	Closed
Expansion Rate	$H_0 > 0$	$H_0 > 0$ (now)
Deceleration Rate	$0 \leq q_0 \leq 1/2$	$q_0 > 1/2$
Density	$\rho \leq \rho_{CRIT}$	$\rho > \rho_{CRIT}$
Hubble Time	$T_0 = \left(\frac{1}{H_0}\right)$	$T_0 = 2/3 \left(\frac{1}{H_0}\right)$

of the expansion rate: $(1/H_0)$. The age of the universe relates to this Hubble time as a $T_0 = (1/H_0)$ for an open universe. In a closed one, however, the mathematics is not as straightforward, yet it has been shown that the age of the universe would be only two-thirds of this value. This means that any handle on the age of the oldest parts of the universe automatically places a limit on whether the universe is open or closed.

Through Hubble's law, Hubble had shown that the universe had a measurable expansion rate, and he tried to find the deceleration rate by extending the Hubble diagram to far greater distances. However, galactic evolution and quasars made it difficult to pin down the deceleration. The CBR, although a result of the embryonic expansion, is in itself a poor tracer of its expansion rate and deceleration. Thus, from the Hubble diagram and the CBR, we do not know if this expansion will go on forever, or lead to cycles of expansion and collapse. We have to consider the cosmological tests—observations used to find cosmological parameters—to find out more about the Big Bang. These cosmological tests have been the driving force of cosmic studies for decades and are where we turn next.

· THREE ·

Cosmological Tests and the Cosmic Questions

WHAT was the universe like billions of years ago? Is it open or closed? What are its next stages, its ultimate fate? These are the questions that have motivated astronomers since Hubble's day. The hunt for those cosmic parameters—H_0, q_0, ρ, and T_0—certainly comprises the key to their search, but these parameters are not what astronomers actually measure. Instead, they must be derived from the relationships that observed quantities make with one another. Below I shall review the general quantities that astronomers observe and describe how they use these quantities to try to extract the values in question. This will also give a perspective on how astronomers were driven to seek out new ways of studying the universe—a new cosmology.

WHAT CAN WE SEE—AND MEASURE?

For all the sophisticated technology that astronomers use to study the universe, the actual number of different things that can be measured is limited. This is partly because the astronomer is a passive scientist—he can only observe celestial objects but cannot, for instance, place a star on a scale and, through experiment, weigh it. Astronomers can see only how this star affects other stars or how its nature may be affected by its mass.

The main things that astronomers can observe are an object's magnitude, angular size and structure, position, and spectra. Motion, when present, can also be sought, as can any changes over time of the intensity, structure, and spectra. One important trick astronomers use is to seek out these quantities over a large range of wavelengths, including, but not limited to, the visible wavelengths. Then, by getting a total picture of how the object "behaves," it is possible to use the known laws of physics, as well as analogies to known properties of matter on Earth, to infer what is happening in celestial objects.

The most severe restriction on astronomers is that they must get nearly all of their information through the electromagnetic radiation they see (light, radio waves, X-rays, infrared, ultraviolet, and so on). So if an object gives off very little radiation, it remains effectively invisible to the astronomer. Compounding this problem is the limited sensitivity of the instruments used today; for example, with photography, there must be enough light present to undertake the desired exposure. This problem of our instruments' sensitivity is one of the principal factors influencing progress on the understanding of the universe. Just as Hubble used a powerful new telescope to discover the Cepheids in spiral galaxies, so more recent astronomers have made their findings thanks to newer, more sophisticated telescopes and devices, some even located in space. But even these inventions have their limitations. (See Plate 2 following page 78.)

Cosmology takes these instruments to their limit, because the distant, "young" universe emits only the faintest measurable radiation. And this is where a troubling problem comes up: What if our inferences about the universe are incomplete or incorrect simply because the equipment with which we make our observations is not sensitive enough and the number of measurable objects is minimal? Is it wise to proceed today—perhaps in the wrong direction—knowing that more and better observations may be available tomorrow?

On the other hand, with pencil and paper (and today, a computer), the theory side of cosmology offers an opportunity to see what physics and mathematics predict should be seen. The path to observing may be difficult, but models guide the observing, directing it toward making those measurements that are most likely to provide insight. Where the predictions of modeling meet the reality of the observations is in the role of the cosmological test. Hubble's law is the result of a cosmological test; LeMaitre's expanding universe model predicted that the receding velocity of a galaxy should be proportional to its distance, a fact that Hubble corroborated through his observations. But to fully appreciate this and other cosmological tests, it is critical that we see what tests have been devised to get a handle on the cosmological quantities, and see what remains to be answered.

COSMOLOGICAL TESTS

Astronomers have settled on the Big Bang as the best model for the universe not arbitrarily but because observations (that is, the recession of galaxies and the CBR) definitely support it; sometimes the support is only crude, and other times it is precise. However, the Hubble diagram is one of many cosmological tests used to estimate the cosmic parameters. As such, these tests are a primary tool of the cosmologist. Here we shall discuss the subtleties of these tests, subtleties that have prolonged the ultimate success of knowing the precise values of the cosmological parameters for over fifty years.

The Hubble Diagram

To recap, Hubble's law came from the relation between the distance of the spiral galaxies versus their receding velocity—a velocity-distance diagram. Originally, using the distance modulus and the magnitudes of Cepheids to infer the distances, Hubble found the distance to about two dozen spirals. Later, he used assumptions about the likeness

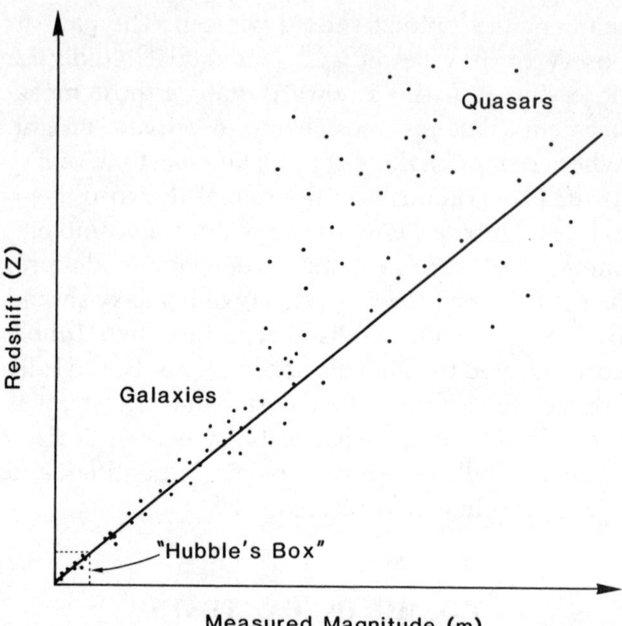

The Hubble diagram, showing the relative scatter for galaxies and quasars. The straight line represents the Hubble law as derived from smaller redshift galaxies. The Hubble box is the region of redshift velocity and measured magnitude explored by Hubble and his collaborators.

of galaxies' absolute magnitude to infer distances, again with the distance modulus.

What astronomers actually observe to construct the Hubble diagram (and Hubble's law) are the spectra of the galaxies and their measured magnitudes (or the magnitude and period of the Cepheids). Then they make an inference—the spectra indicate velocities, via the Doppler effect, and the measured magnitudes imply distances, via the distance modulus.

The Big Bang is a cosmological model that predicts an expanding universe, and the expansion yields the velocities of recession that we infer from the observations of the spectra. The fact that the velocities increase directly as the distance does is the clincher for the Big Bang. If nearby galaxies moved as the fifth power of the distance, for example, the Big Bang wouldn't be viable. The Hubble constant—derived from the slope of the line connecting the data points for many different galaxies—is the expansion rate, as predicted by the theory of the Big Bang. Meanwhile, the deceleration parameter should also show up on the Hubble diagram, in the form of a curved extension of the straight line of Hubble's law for dim, highly redshifted galaxies. The degree of the curving corresponds to differing amounts of curvature for the universe.

Even with possible effects of galactic evolution, the situation of extending the Hubble diagram to very great distances is complicated by this curvature and another strange but accountable effect—that of a relativity correction to the redshift.

The redshift correction comes up from straight common sense: If Hubble's law continues as a straight line indefinitely, then at what distance will the velocity of recession of a galaxy be greater than the speed of light? For example, if the value of Hubble's constant was 75 km/s/megaparsec, then Hubble's law, $V = H_0D$, predicts a faster-than-light redshifted velocity at 4,000 megaparsecs—distant, but not extremely so. Einstein showed, in his special version of relativity, that very fast-moving objects undergo a distortion of dimensions compared with those of the observer; time moves slower, and so on. This means that spectral shifts of the receding galaxies must be corrected for the relativity effect. The correction comes about automatically by defining the Doppler effect shifting of the spectrum as a *redshift term*, Z, which is a representation of how much the shift has moved in wavelength compared with the wavelength of the spectral line (or lines). Effectively, astronomers stopped discussing redshifted velocities, replacing them with the more elegant redshift, Z. This means that Z can be defined by the velocity of recession as

$$Z = \left(\frac{1 + V/c}{1 - V/c}\right)^{1/2} - 1$$

where c is the speed of light (about 300,000 km/s). When V is much smaller than c, $Z = V/c$. Expressed in this way, Hubble's law becomes $cZ = H_0D$. The deceleration parameter is a way of describing the

As an example of the relativity correction for very high speeds, an observer (a) compares the observed speed of a passing bullet with its actual speed. The difference is shown in the plot at (b). The bullet seems to be moving faster than the speed of light, although its true speed is smaller than the light speed limit by just a fraction.

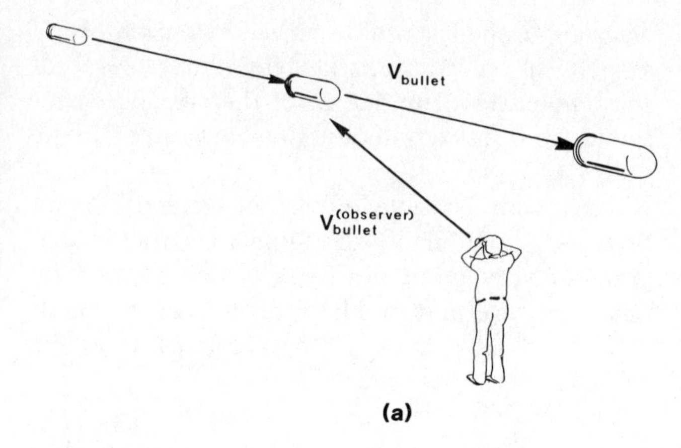

(a)

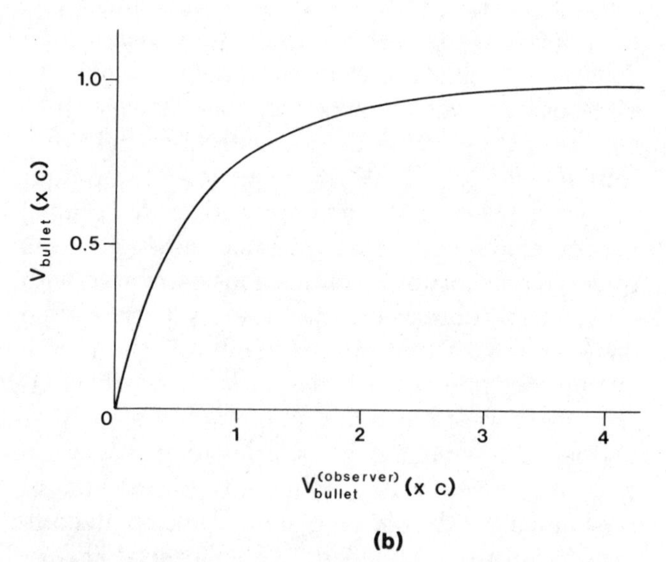

(b)

curvature of the universe through the change in Hubble's constant over time. Only at very great distances will the effects of q_0 be seen. So when the relativity effects are corrected for in the redshifts and measured magnitudes, the straight line of Hubble's law becomes curved at large redshifts. More accurately, the redshift-distance relationship, including the deceleration factor, gives the following.

$$cZ\left(1 - \frac{Z}{2}(1 - q_0)\right) = H_0 D$$

Even when all these effects are understood and applied, there is the overriding problem of galactic evolution. This enters the Hubble diagram through a change in galaxies' magnitude over time—it is an effect unrelated to the expansion of the universe or its curvature. The magnitudes measured for galaxies are the sum of distance effects and galactic evolution, and there is no simple way of separating the two.

The significant effect of galactic evolution has become accepted only within the last 15 years. The New Zealander-American Beatrice Tinsley, after many years of well-founded argument, convinced her colleagues in the mid-1970s that earlier studies of the Hubble diagram and other cosmological tests were severely contaminated with the effects of galactic evolution and required new lines of pursuit. This corroborated and enforced earlier notions of these ideas, first described by Walter Baade.

Because the farthest galaxies are the ones that trace the universe's curvature, and these farthest galaxies are the ones whose youth suggests the greatest difference in galactic evolution, solving for q_0 from the Hubble diagram is virtually impossible. Solving for H_0, however, is a bit more reliable, albeit subject to uncertainty.

When Hubble derived the Hubble law, his first estimates were greater than 500 km/s/megaparsec.

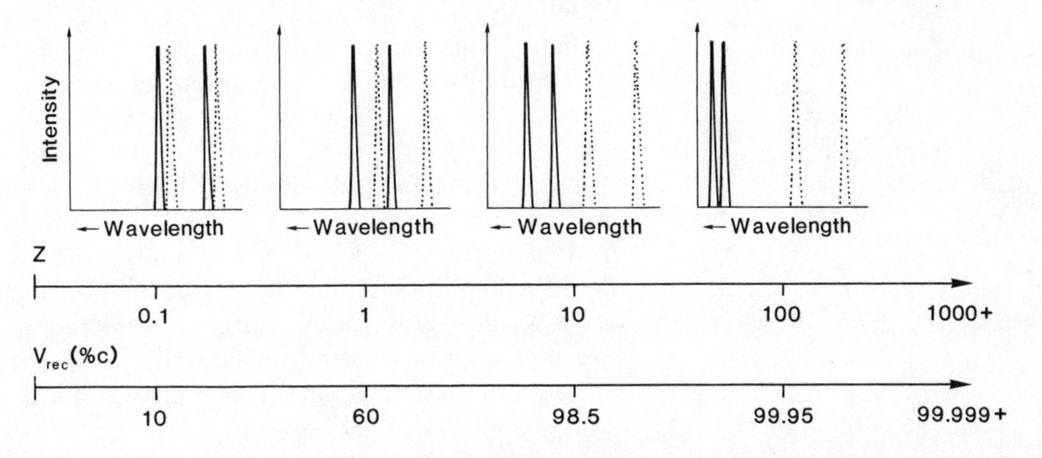

The relation between spectral shift for two spectral lines, their redshift, Z, and their redshifted velocity of recession as a percent of the speed of light. The dotted lines represent the lines' wavelengths for zero velocity and zero Z.

By the time of his death in the late 1950s, he had reduced the estimate to less than 200 km/s/megaparsec (thereby increasing the derived age of the universe by more than two times). Hubble wasn't being flaky but finicky; the uncertainty of the measurements of galactic magnitudes, when properly understood, allowed him to revise his estimate with the proper amount of caution. Today, more-reliable values of the magnitudes of distant galaxies assign the value of H_0 between 40 and 100 km/s/megaparsecs. Later we shall discuss more contemporary estimates.

The Hubble diagram, with all its problems, has continued to be the main motivator for astronomers to study the properties of the expanding universe. They often speak of calibrating Hubble's constant. What this means is that they will use other methods to find the distance to galaxies and from the redshift and distance information derive H_0. I shall describe these other distance yardsticks shortly, as well as their success in terms of the values of H_0 they suggest.

The principal reason the Hubble diagram is important is that it demonstrates that the universe is expanding. We have yet to know the precise rate of that expansion, or the change in the expansion rate, based purely on the Hubble diagram. Despite its problems, the diagram remains one of the main tests that cosmologists use. Perhaps it might ultimately reveal H_0 and q_0 with great accuracy, with sufficient understanding of galactic evolution and improved observations of the distant galaxies.

Angular Size—Redshift Test

In 1959, British astronomer and mathematician Fred Hoyle proposed that the curvature of the universe would have a strange effect on the angular size of the most distant galaxies. He found that our normal notion that an object's angular size de-

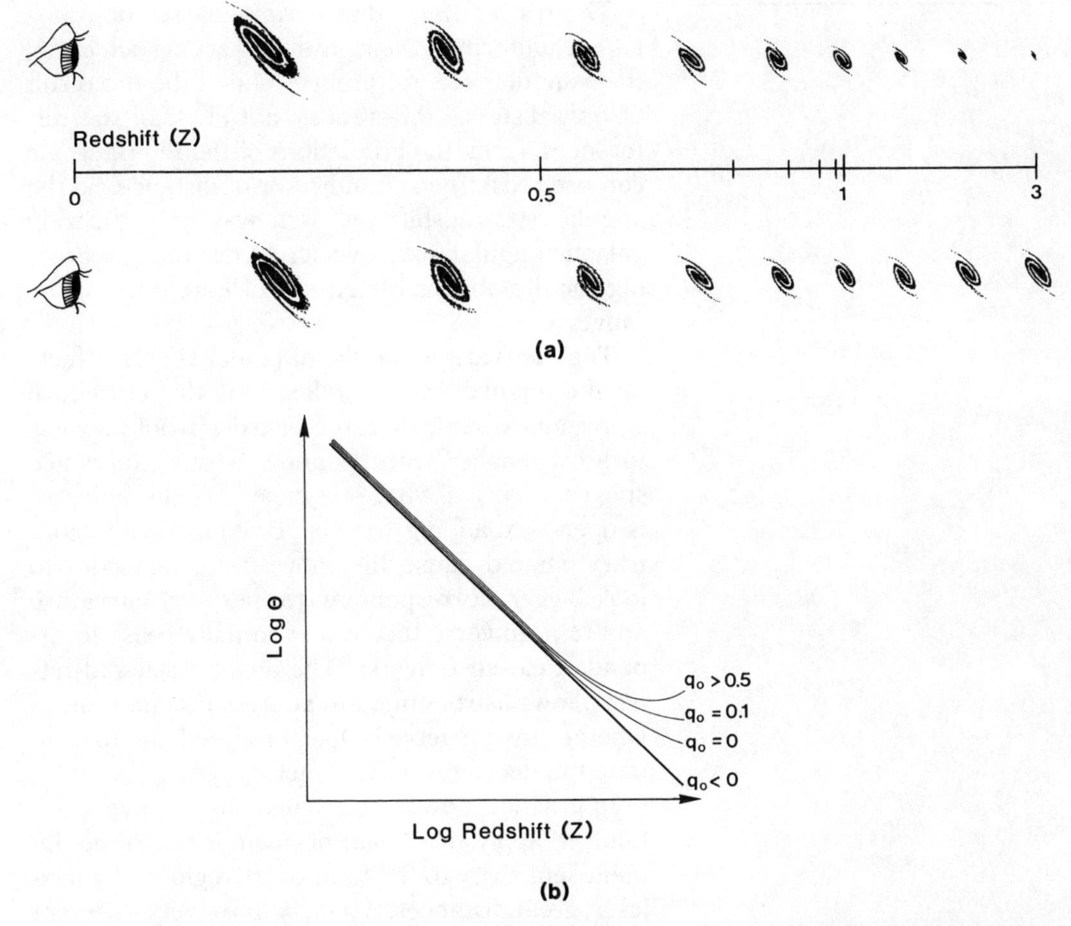

The angular size-redshift test, showing the effects of the universe's possible curvature on the angular size (a), and a plot that shows how the sizes would change for different curvature models (b).

creases in proportion to its distance might not be true over the large span of the universe, and that angular size could, in principle, be used to find q_0.

Space, as portrayed in the Big Bang model, may be curved across the universe, a concept that is unlike our earthly experience of flat, or Euclidean, space. This curvature causes distant galaxies to shrink in angular size with increasing distance—up to a point. Then the galaxy actually looks bigger in size from the curvature of space's distortion of it. Assuming, as demonstrated by the Hubble diagram, that redshift can be used to infer distance, it is possible to design a cosmological test that searches for this odd effect.

The test assumes that certain classes of galaxies are uniform in their physical size. In such cases, their angular size differences should be the result of only distance differences, not physical size differences. From the predictions of the Big Bang, we can use redshift as an indicator of distance. So the angular size-redshift test is a way of comparing galactic angular sizes over great distances, searching for the observable effects of curvature of the universe.

That curvature should have noticeable effects on the angular size of galaxies. If this curvature were nonexistent, then the galaxies would appear to grow smaller with distance, as our experience suggests. This is what one expects if the universe is open—expanding forever. Braking, on the contrary, should cause the more distant galaxies to look bigger, corresponding to "positive" curvature and to a universe that will eventually cease to expand: a closed universe. The angular size-redshift test allows astronomers to address the question of whether the universe is open or closed and to evaluate the deceleration parameter, q_0.

In practice, however, this test has been very difficult to apply. The main problem is one of equipment sensitivity to the faint outer regions of galaxies at great distances. A galaxy looks very different under varying time exposures because the small core is much brighter than the more extended, outer structure (spiral arms in spiral galaxies, for example). Similarly, if a galaxy is faint to begin with, its outer region will probably not be visible at all. Measuring the angular size of the galaxy's image will be a misleading enterprise if you can't be sure you're seeing the whole thing. The issue of galactic evolution, again, throws yet another wrench into the works. What if younger galaxies have much brighter cores, compared with their extended region, than younger ones? Is it really meaningful to attribute to all galaxies of one type the same physical size?

These problems have made the angular size-redshift test an as yet unrealized way of finding q_0. However, ongoing work seeks to extend our sensitivity to the faint outer parts of galaxies and to account for the galactic evolution in a way that can indicate the angular size increase at large redshifts. The angular size-redshift test may prove an effective one years down the road.

Source Counts Test

Taking for granted our Big Bang model of a still-expanding universe, all the galaxies should become increasingly spread out from each other with time because of the volume change of the expanding universe. But for given increments of time, and thus distance and redshift, there should be a relation between this expansion and how many galaxies you see within each increment. This method of counting galaxies and "binning" them according to their redshifts is called the *source counts test* and is a way of solving for q_0.

For our observation of very distant galaxies, we can assume that dimness is an index of distance,

The source counts text—early in the universe, galaxies may be more densely packed than they are now.

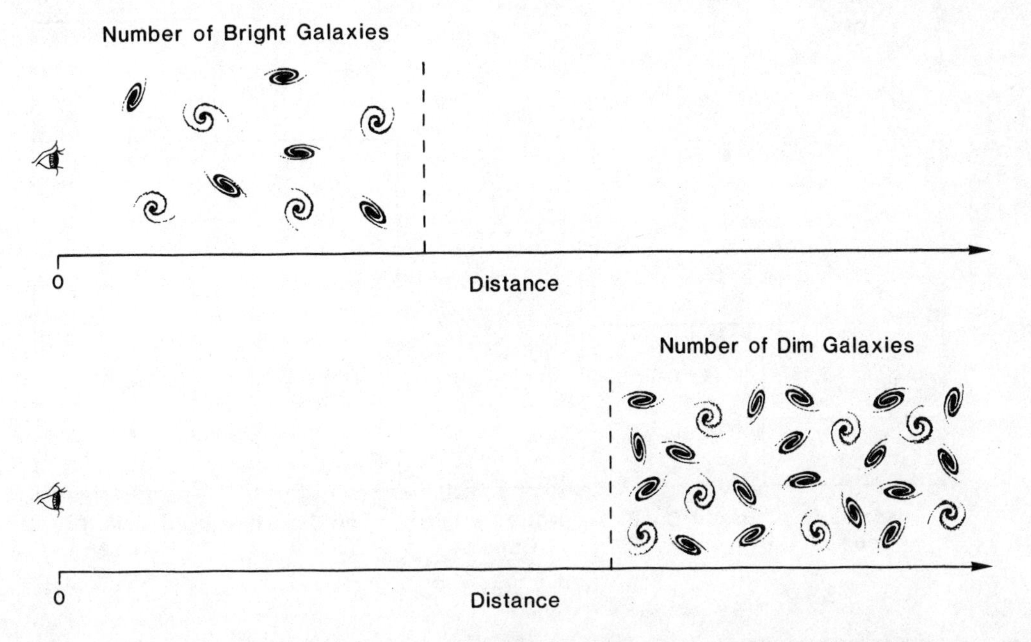

through the use of the distance modulus. Thus another type of source counts test, less precise but easier to do (because spectra are much harder to measure) than the number-redshift version of the source counts test, uses the measured magnitudes of galaxies, incremented over small ranges of magnitude, to do the source counts.

If the universe were not expanding, then the number of galaxies counted per increment would reveal a characteristic shape, whereas various val-

In a universe of candles (a), we could count the number of candles at increments of magnitude or redshift and learn something about the nature of this candle universe (b). The source counts test is a similar method.

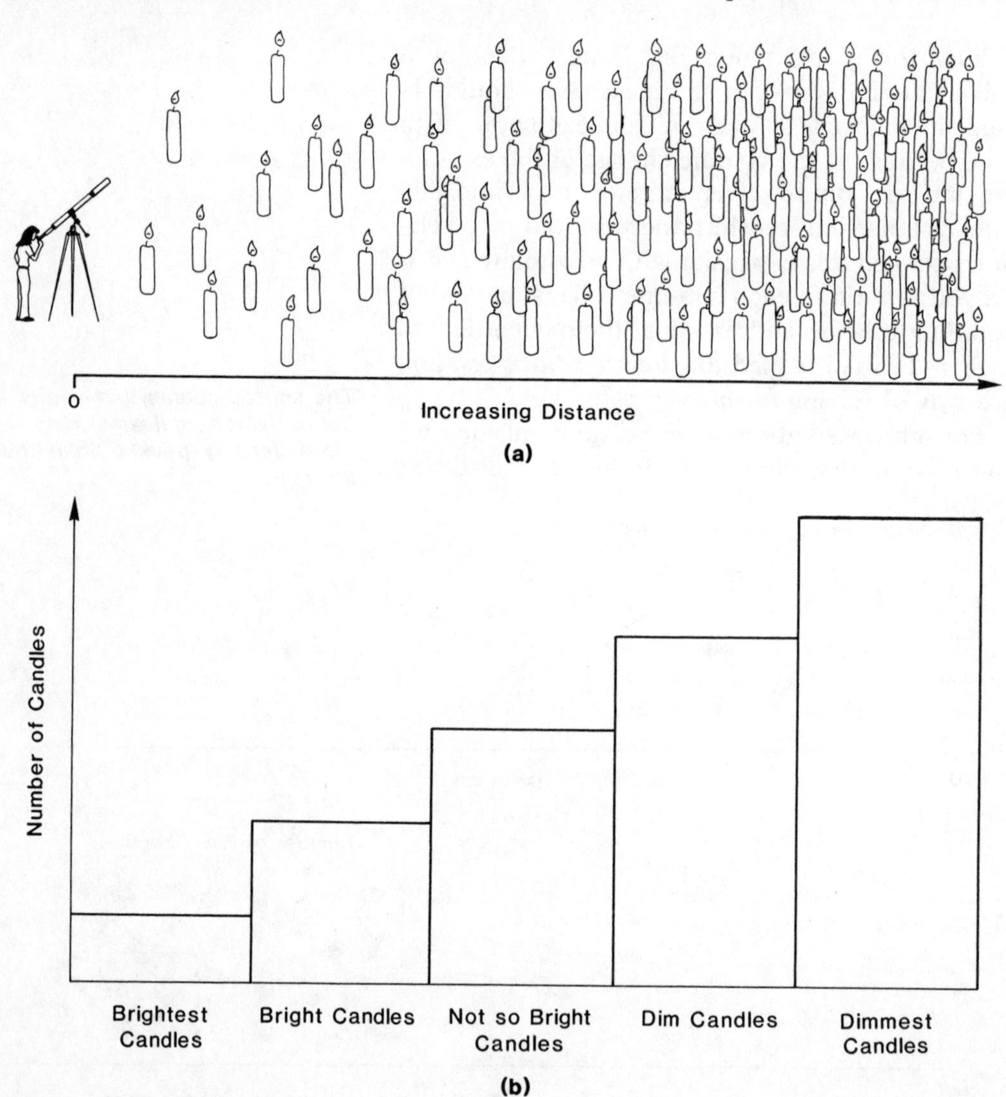

ues of expansion and deceleration would reveal varying shapes.

Hubble was among the first to use the source counts test; in fact, he considered its use as important for cosmology as the Hubble diagram. The reason for this is that he thought it would be easy to show that the universe did not follow the fixed curve of a nonmoving, or static, universe in which the number of galaxies per magnitude increment would be described by the equation $N(m) = 10^{0.6\Delta m}$ where $N(m)$ is a shorthand for the number of galaxies counted with a magnitude lying within the increment Δm.

Hubble did indeed show that the actual observations did not follow this equation, and he concluded that the value of q_0 must indicate a very closed universe. Yet if the universe were very closed, it might now be contracting, despite the observation of expansion.

The problem was that Hubble was not aware of the subtle factors that had to be applied to his magnitudes; certain correction factors were needed, just as with the Hubble diagram. So his conclusion—that the universe was very closed—was not correct.

In 1961 the American astronomer Allan Sandage, a colleague of Hubble's, made an important, and troubling, insight into the source counts test. He showed that the version that used magnitude increments (rather than redshifts) could never really say anything about q_0 because there had to be a correction to magnitudes based on the relativity effects—and this canceled any observed effect from the curvature effect.

The main problem with the source counts test is that the very dimmest galaxies hardly appear on photographs. It is impossible to get their spectra, which require even greater sensitivity. Only a fraction of the dimmest galaxies can have their spectra taken, while many more can be measured for their magnitudes. This led astronomers to be wary of a selection effect, a bias where the astronomer chooses only those galaxies for which he or she can measure spectra, rather than deciding on a choice of galaxies and getting all their spectra. Concerned about coming to the wrong conclusions regarding q_0, astronomers decided to wait—until present times—to use the source counts test in order to use much more sensitive instruments to remove the selection effect.

The Distance Calibrators

A handful of methods exist that, in principle, give more accurate estimates of distance than the use of galaxy magnitudes. These methods are usually applicable only to a very few galaxies, so they are not usually called cosmological tests; rather they are called *distance calibrators.* Below I shall discuss a few of these.

When Henry Curtis found novae in the spiral nebulae, it led him to believe that the spirals were outside of our own galaxy. Even novae, however, are too dim to see in all but the closest spirals.

Yet there is another type of exploding star, similar to a nova, that can be used to calculate distances. These are exploding stars called *supernovae,* giant red stars that suddenly collapse upon themselves, burning up millions of years worth of hydrogen fuel in a few hours. Supernovae are extremely rare (fortunately—it is more healthy to be far from a supernova than close to one) and happen only once every few hundred years, typically, in a galaxy of a thousand billion stars. When observed in very distant galaxies, however, supernovae do prove useful in calculating distances to those galaxies. (See Plate 10 following page 78.)

The technique of calculation is reminiscent of the Cepheid yardstick. Supernovae brighten quickly (sometimes by more than ten magnitudes), followed by a period of slow dimming, which may take years to return to the pre-supernova magni-

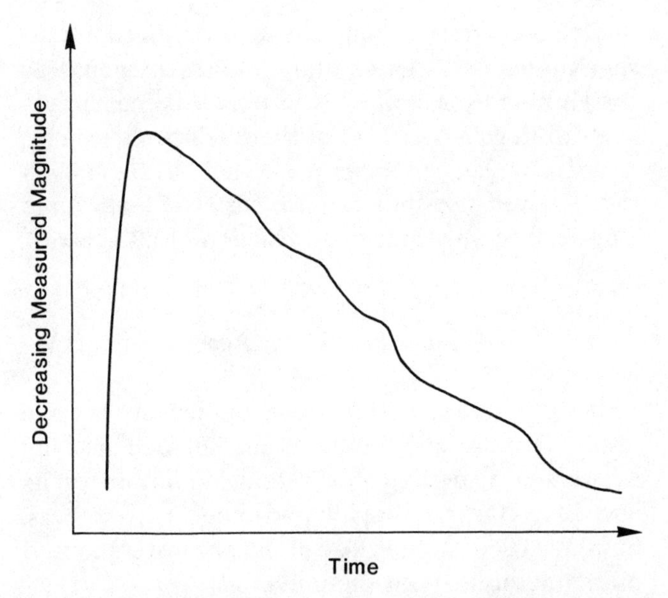

Light curve of a supernova. The way the peak magnitude decreases with time is a signature of the supernova's absolute magnitude.

tude. It is believed that the slight differences in the time for brightening and dimming break supernovae into a few different groups, and that for each group a distinct absolute magnitude is reached when the supernova is at its brightest. Hence the brightness change can be used to infer an absolute magnitude, from which the distance modulus yields a distance. With the distance a known quantity, it is possible to measure the redshift via the spectra of the parent galaxy of the supernova, and from that Hubble's constant is derived: $H_0 = cZ/D$.

Another calibrator method uses the size of gas clouds, called *HII regions,* to measure distances. These gas clouds are far more tenuous than any terrestrial cloud and are mostly made up of hydrogen. But they glow from the starlight of embedded young stars, and they give off spectral lines of hot hydrogen. Such HII regions in our galaxy are often hundreds of parsecs or more across. In general, there is a tendency for the brightest HII regions in galaxies to have the biggest sizes, and the range of this biggest size is fairly small. So by measuring their angular sizes, we can use trigonometry to infer the distance to the parent galaxies.

These two techniques have been used to assign values between 45 and 75 km/s/megaparsec for Hubble's constant. The problem is that only a few dozen galaxies, most of them relatively nearby, have been used for supernovae and HII region calibration because they are not easy to identify and can be seen only in a small fraction of all galaxies.

Globular Cluster Ages

Although there is no direct way to measure the age of the universe, astronomer Allan Sandage has long advocated a method that sets a minimum for its age. Like Harlow Shapley and other astronomers, Sandage used our galaxy's globular clusters to find out something about the universe.

The colors and the spectral lines of stars in globular clusters indicate that those stars have undergone a different growth pattern from other stars, such as the sun. Apparently, the globular cluster stars burned up their hydrogen at a different rate from other stars. By measuring the amount of material that is a by-product of this burning—helium and other elements—it is possible to set a limit to the age of these stars. Using this method, Sandage found that these globular cluster stars are very old—between 12 billion and 18 billion years of age. So, by assuming that the universe is older than any of its individual parts, we might infer that the universe is at least 12 billion years old, and perhaps much older.

A minimum age for the universe helps resolve whether it is open or closed. In a closed universe, the age is equal to two-thirds of the inverse of the value of Hubble's constant. From the other cosmological tests, it appears that H_0 lies between 45 and 75 km/s/megaparsec. Then we can use Hubble's constant to derive the Hubble time and compare it with the minimum age, as shown in Table 3.1.

Table 3.1 Hubble's constant and the age of the universe, T_0

H_0	Age (closed)	Age (open)
45	14 billion years	22 billion years
75	8 billion years	13 billion years

Notice that the 8 billion years of $H_0 = 75$ km/s/megaparsec in a closed universe does not fall within the minimum age limit. Here lies an interesting prediction: If H_0 is found to be anything more than about 55 km/s/megaparsec, the universe must be open. Clearly, Sandage's estimate of minimum age, combined with the Hubble constant, makes for a clever and powerful argument about the openness or closeness of the universe.

The Density of the Universe

If most of the mass of the universe lies in its galaxies, and if we could weigh a galaxy, then we could extrapolate the density of matter in the universe. And if we knew that density, a comparison of its value to the critical density (the density at which the universe becomes closed) could reveal whether the universe is open or closed.

Weighing a galaxy—getting its mass—is a difficult and indirect task, based on assumptions and inferences. The most common approach is to invoke an assumption astronomers make about the motion versus potential energy (energy that could be turned into motion) of a galaxy lying in a grouping, or cluster, of companion galaxies. Then an approach called the virial theorem can be used.

The virial theorem describes a simple idea. It states that for a galaxy to be part of a galactic cluster, the galaxy must be "orbiting" the mass-center of the cluster at a particular velocity that offsets the galaxy's gravitational attraction to its neighbors. Otherwise, the galaxy would disperse, losing

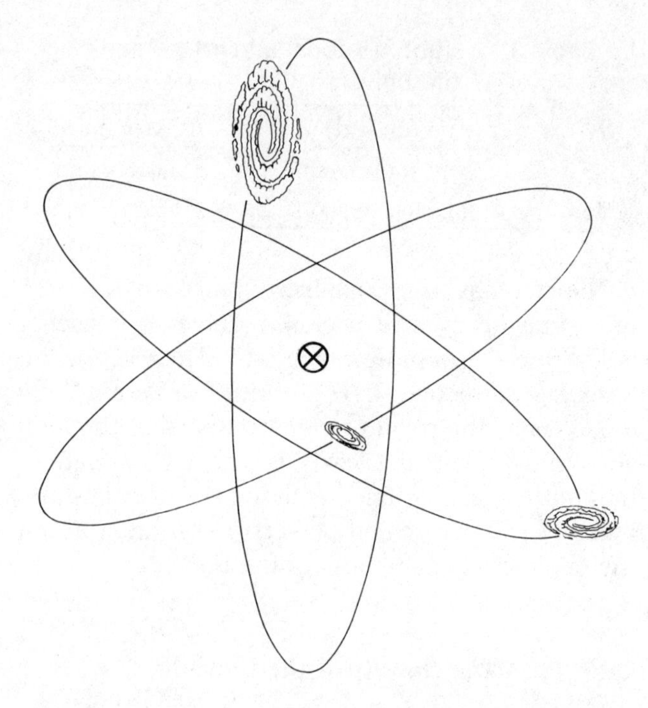

The virial theorem—here three galaxies in this cluster move around the center of mass. The speed at which each moves tells its mass.

its parts to other galaxies. Although this breakup happens quite commonly (see Chapter 5), there are many galaxies, such as our Milky Way, that have balanced off gravitational attraction of neighbors by motion.

Even a single orbit is much too long to observe—it would take many millions of years—but the motion can be extracted by looking at the spectra of the galaxies in the cluster. With spectra, the relative velocities of the galaxies, through the Doppler effect, quantify how much energy is tied up in motion, while gravitational attraction balances it out. The gravitational attraction depends on the mass of the galaxy and its distance from other members, so by looking at the motion of all the members, it is possible to infer how much mass the galaxy must have to execute this delicate balance. Not all galaxies yield their masses via the virial theorem, but typical values are between a billion and a thousand billion stars of the mass of the sun.

To compute the mass of the universe, we can

scale up the mass of the galaxies times the number of likely galaxies in the universe. This naive method is about as good as we have; currently, a more direct way of getting the mass of the universe doesn't exist. Not surprisingly, the number, expressed as a density compared with ρ_{CRIT}, is uncertain by a factor of three or more. It lies between 5% and 15% of the ρ_{CRIT}, strongly suggesting that the universe is open—if we see all the universe.

The most questionable feature of this type of extrapolation is the comparison of one galaxy of known mass to another. For example, if the galaxy of known mass has a certain absolute magnitude, it becomes convenient to speak of a mass-to-light ratio, or M/L, which relates the amount of mass as compared with the amount of light. The sun is defined as having a M/L of 1; a M/L of 1,000, for example, indicates a star that is extremely dim, considering its mass, in comparison to the sun. A galaxy with a M/L of 1,000 would comprise a huge collection of extraordinarily dim stars. There is some startling evidence that indicates that much of the mass of a galaxy may not give off sufficient light for us to see it. Can we merely take the light from galaxies and assume there is a quantity of mass that it corresponds to?

Perhaps much of the mass of the universe lies hidden from us, too dim to see. Astronomers call this the *dark matter problem,* or, alternatively, the missing mass problem—*missing* referring to the lack of apparent mass to close the universe.

SPECIAL PROBLEMS—QUASARS AND THE EARLY UNIVERSE

All cosmological tests depend on galaxies to act as markers of the universe's course. Unfortunately, at the greatest redshifts, galaxies are rare and quasars common. There are no very close quasars and few very distant galaxies. Yet quasars are hard to

Plate 1 The Caracol ruins, an ancient Maya observatory at Chichén Itzá, Yucátan, Mexico. (*Courtesy E. C. Krupp, Griffith Observatory.*)

Plate 2 Astronomer Donna Weistrop prepares a modern telescope for the evening's spectroscopic observations of quasars. (*Courtesy NOAO.*)

Plate 3 Sunset at the VLA in New Mexico.

Plate 4 False-color radio photograph of a galaxy jet. The core of the galaxy is at the lower left. (*Courtesy NRAO.*)

Plate 5 This heavily enhanced false-color photo shows a trail emanating from a galaxy (right) to a quasar (left), MK 205. Is this a bridge between a galaxy and a quasar, or is it a jet of the more-distant quasar projected against the foreground galaxy? (*Courtesy NOAO.*)

Plate 6 The visible image of this galaxy has been blocked out to reveal the much dimmer visible-light jets, here shown as spikes. (*Courtesy H. C. Arp.*)

Plate 7 A composite photograph of a cluster of galaxies, with visible jets (blue and pink) and radio lobes (green). (*Courtesy NRAO.*)

Captions continue on page 79.

Plate 1

Plate 2

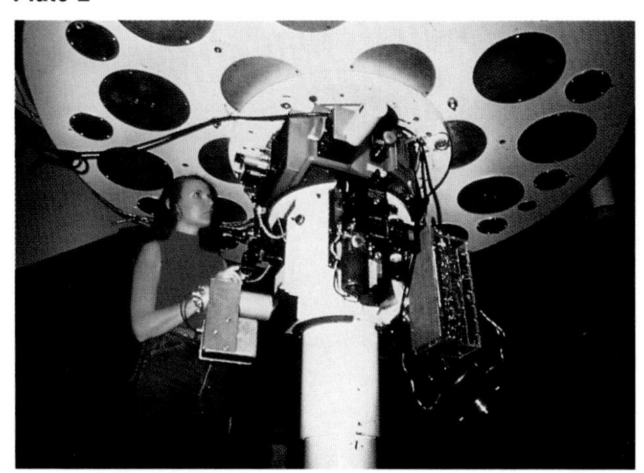

Plate 3

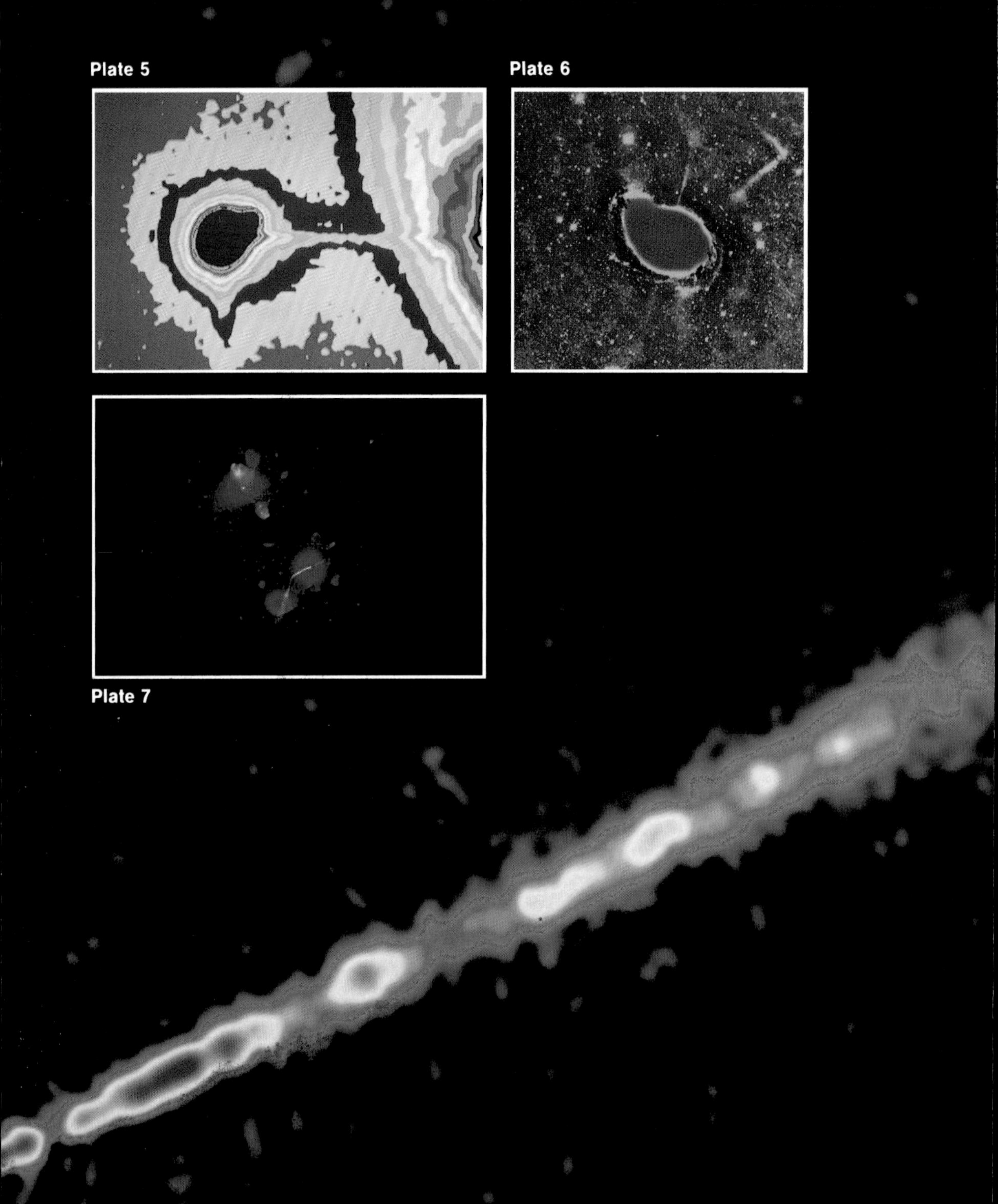
Plate 5
Plate 6
Plate 7

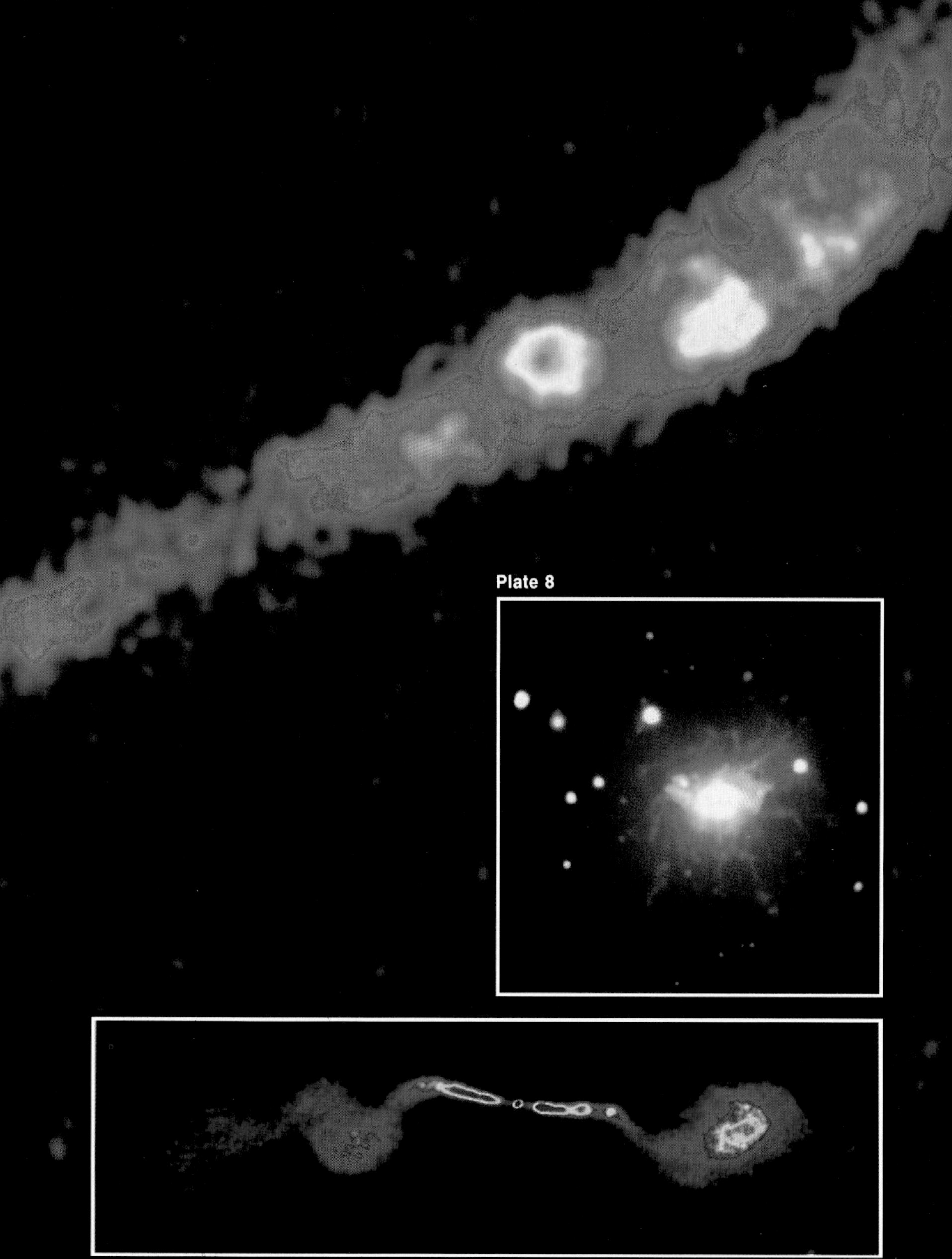

Plate 8

Plate 9

Plate 10

Plate 11

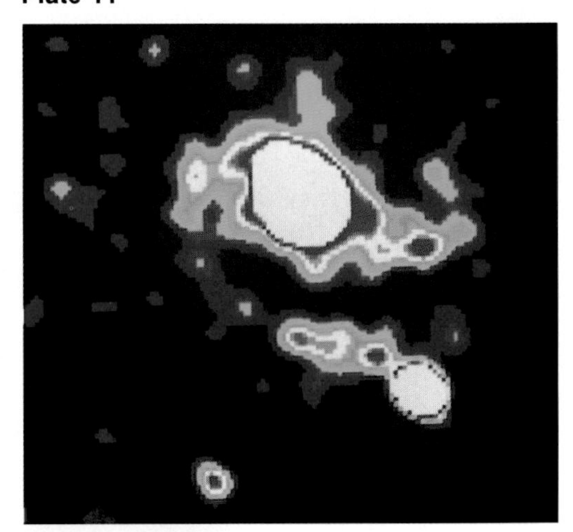

Plate 12

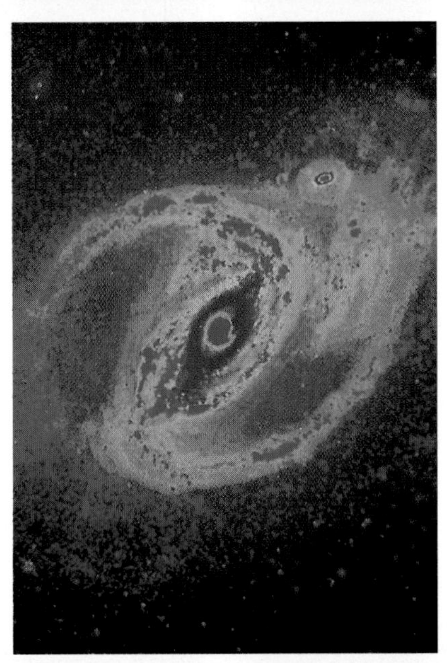

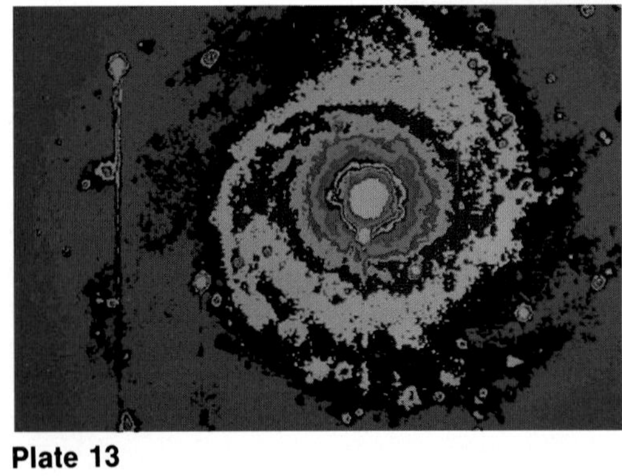

Plate 13

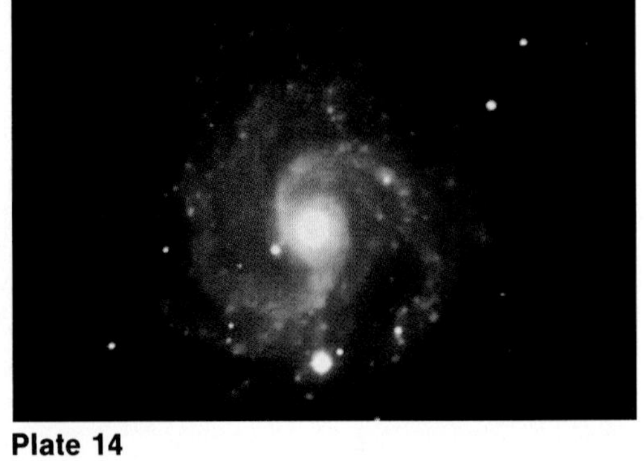

Plate 14

Plate 8 The active galaxy 3C84, with filaments of gas coming from its violent core. (*Courtesy NOAO.*)

Plate 9 Jets and lobes of a DRS in false color. The outer lobes have been distorted by the slight sideward force of the intergalactic medium. (*Courtesy NRAO.*)

Plate 10 A supernova (center, right), fleetingly as bright as the core of the galaxy containing it. (*Courtesy R. Schild, SAO.*)

Plate 11 False-color radio photograph of a feebly emitting radio galaxy and its new supernova (below), an individual radio source. (*Courtesy Nobert Bartel.*)

Plate 12 Color-enhanced view of a barred spiral; compare this with the "blocked" photograph of the same galaxy in Plate 6. (*Courtesy H. C. Arp.*)

Plate 13 False-color photograph of a galaxy, with part of its cocoon apparent as the very dim, purple region. (*Courtesy NOAO.*)

Plate 14 A spiral galaxy, in true colors. (*Courtesy R. Schild, SAO.*)

place into the usual scheme of cosmological tests because they are so unlike galaxies (unless they themselves are strange galaxies). The realm of the quasars, and even of the most-distant galaxies, is a relative realm of the unknown.

Considering the uncertainty surrounding quasars, it's not too surprising that some astronomers question whether they are very distant objects. Halton Arp, an American astronomer, has long defied the existing convention that quasars are like the galaxies. He finds that there is a tendency to have too many quasars lying very close in angular proximity to galaxies, even when those galaxies have very different redshifts from the quasars. This may be a projection of nearer galaxies against

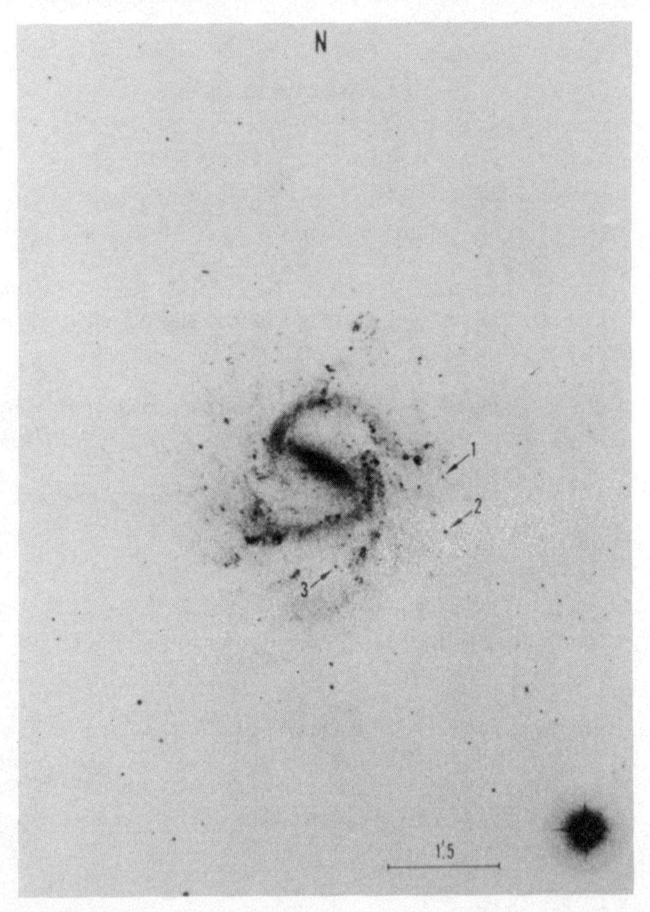

The different appearances of a galaxy and quasar. The arrows point to quasars in close angular proximity to a galaxy of far greater angular size. It has been suggested that quasars may be companions to this galaxy, despite the huge differences in redshifts between the galaxy and quasars. (Courtesy H. C. Arp.)

distant quasars, but Arp believes that too many quasar-galaxy associations exist to be explained by this chance projection. Because astronomers tend to equate redshift with distance (as in Hubble's law), this notion questions the fundamental assumption of how we scale the universe's distance—are quasars all shot out from galaxies and not at the distance their redshifts imply? This redshift controversy, of importance in the mid-1970s, has since become less of an issue now that several thousand quasars, some in clusters, have been found. Chance projection seems much higher now than suggested by the probabilities derived earlier, which were computed with less knowledge of quasar numbers. There are some cases, however, where quasars of great redshift appear to be connected to nearby galaxies—enigmas yet to be understood. Until (if ever) the evidence for quasars attached to galaxies becomes overwhelming, the idea of quasars as offshoots of galaxies, as Arp suggests, will be the underdog of theories on the nature of quasars. If, however, that day comes, our ideas about the universe would be revised radically. (See Plate 5 following page 78.)

UNANSWERED QUESTIONS

As of the late 1970s, despite decades of intensive work by theorists and observers, few questions about the nature of the universe had been answered. The solid conclusions were these:

1. The universe is expanding at a rate, H_0, somewhere between 45 and 75 km/s/megaparsec.
2. The discovery of the CBR, combined with our knowledge about expansion, indicates that our universe originated as a colossal explosion in a fixed point in time.
3. The universe cannot be younger than 12 billion years.

Table 3.2 Cosmological tests and cosmological parameters

Cosmological Test	Characteristics Measured	Cosmological Parameters
Hubble Diagram	magnitude, redshift (Z)	H_0, q_0
Angular Size-Redshift	galaxy angular size, Z	q_0
Source Counts	# of galaxies, Z interval	q_0
Distance Calibrators	magnitude, angular size, Z	H_0
Globular Cluster Ages	star colors, spectral lines	T_0
Density	virial mass from motion, M/L	ρ

4. Present values of the density of the universe are only a small fraction of the critical density; the universe must be open if we are to believe that the visible matter in galaxies indicates how much matter the universe contains.

The unanswered questions can be summarized as follows:

1. What is the nature of the quasars? What can we learn about the early universe from them? Are there other observational things that can tell us more about the early universe?
2. When we look at galaxies, how much of their mass is invisible (dark matter)? Can we find ways of getting the mass of galaxies other than extrapolating their mass from their light?
3. How might the process of galactic evolution be understood? Can we ever be able to model the differences between similar galaxies of very different observed ages?
4. Is the universe open or closed? What are the values of the cosmological parameters?

5. What happened to the universe in its earliest stages? How did galaxies form, and how does this relate to the Big Bang?
6. What will the universe be like billions of years from now?

In Part 2, we shall see how the new findings in cosmology have addressed these questions and how our knowledge about the universe has changed in a few short years—an infinitesimal fraction of the universe's age.

TWO

NEW VIEWS

· FOUR ·

The Radio Universe

HIDDEN beneath the glare of the visible universe lies another view, unique and insightful, fascinating and bizarre. It is a radio universe, in which cool and hot gases, magnetic fields, and atoms give us knowledge on the structure of galaxies, limits to the amount of invisible matter, and the changing conditions of the universe through past time.

In 1932 the science of radio astronomy got off to an uneventful start when a Bell telephone radio engineer, Karl Jansky, discovered that the extra hiss he was finding in a radiotelephone linkup was coming from beyond Earth. But it wasn't until the mid-1940s that astronomers, using radar hardware left over from World War II, began pursuing this isolated discovery in earnest.

The first surprise the radio astronomers encountered was the reason the celestial hiss was there in the first place. They knew that radiation is released by electrons when they are heated; the hotter the electrons are, the more energy they release in the shorter wavelengths of the electromagnetic spectrum, such as light, ultraviolet, and X-ray. This thermal radiation should be very feeble at radio wavelengths, since stars are so hot that they emit most of their radiation in visible light. The relative intensity of the radiation at different wavelengths has a unique spectral signature that indicates the temperature of a celestial object, such as a star or planet. But the radio emission that these astronomers were detecting from the sky lacked the spectral signature of a thermal radiator. And by 1950, this odd radio emission had been picked up from dozens of directions in space.

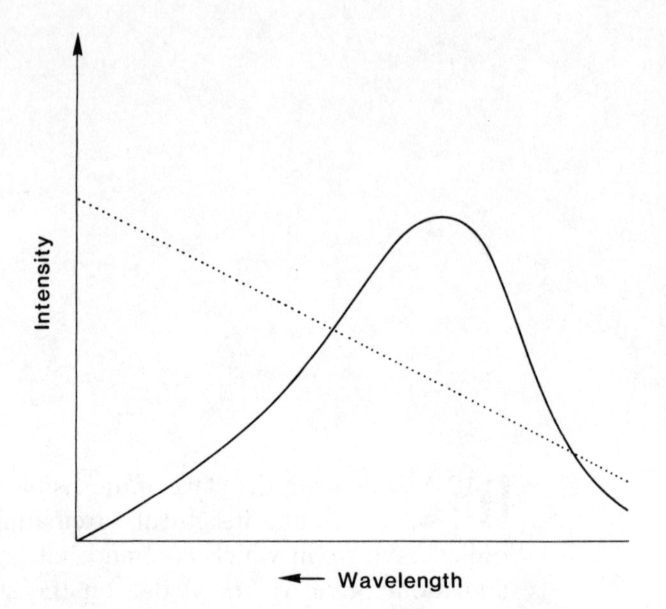

Continuum spectrum for a thermal radiator—a hot object (solid line)—and a synchrotron source (dotted line).

The solution to this paradox came out of nuclear physics. When very energetic electrons were placed in a magnetic field, they were guided by the magnetic lines of force. The electrons spiraled in the field, constantly emitting radiation. This mechanism of radiation is called *synchrotron radiation,* after the particle accelerator from which it was first found. In 1952, Soviet astrophysicist Iosef Shklovskii pointed out that the synchrotron radiation was similar in characteristics to the odd radiation of cosmic radio sources. Thus the sources must indicate "stars" where vast numbers of electrons spiral in weak but widespread magnetic fields. Yet they could not be stars in the normal sense, since stars give off nearly all their light from thermal radiation. A major question remained unanswered: What were the cosmic radio sources?

Cygnus A (not to be confused with the parallactic star 61 Cyg A)—the designation for the strongest cosmic radio source—gave the first hint to the nature of the cosmic radio sources. Upon study with state-of-the-art antennae, Cygnus A was distinguished as two separate, but related, radio sources lying on opposite sides of a galaxy. But

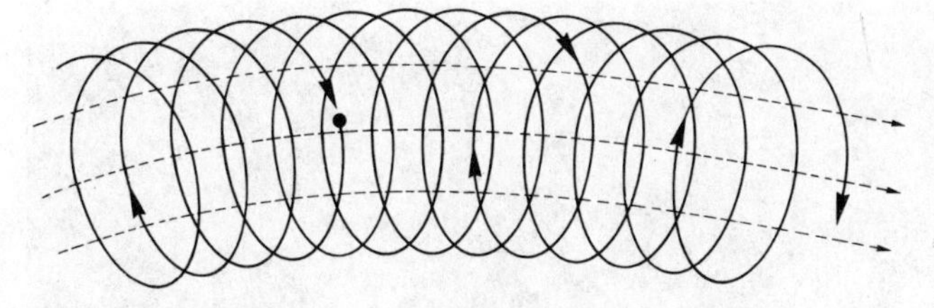

An electron spirals when embedded in a magnetic field, causing it to radiate energy.

these "lobes" lay far outside the visible regions of the galaxy and so did not appear directly part of it. Soon, many other cosmic radio sources were seen to have this double-lobed structure, whose range almost always dwarfed the visible extent of the parent galaxy. The double structure gave rise to the name *double radio sources* (DRS's).

INTERFEROMETRY AND THE DRS'S

What phenomenon causes the DRS's? How do the spiraling electrons get to each lobe of a DRS? How does the parent galaxy control the lobes? These were the questions that Cygnus A and other such objects inspired.

To answer them, better views, with finer detail of the DRS's, had to be obtained. This depended primarily on the resolving ability—viewing of details of very small angular sizes—of the radio dish or optical mirror. The best way to get good resolution is to build a very large dish or mirror. But this becomes financially impractical for radio observations because the resolving power depends not only on the size of a dish or mirror but also on the ratio of wavelength to that size. Since radio wavelengths are a million times longer than visible light, a half-meter optical mirror has the same resolving power as a (hypothetical) radio dish some 500 kilometers in diameter! In a practical sense, when it comes to resolving power, the human eye beats the biggest single radio telescope (the 305-meter behemoth in Arecibo, Puerto Rico) hands down. Fortunately, a clever shortcut offers a cheap way to get excellent resolving power at radio wavelengths.

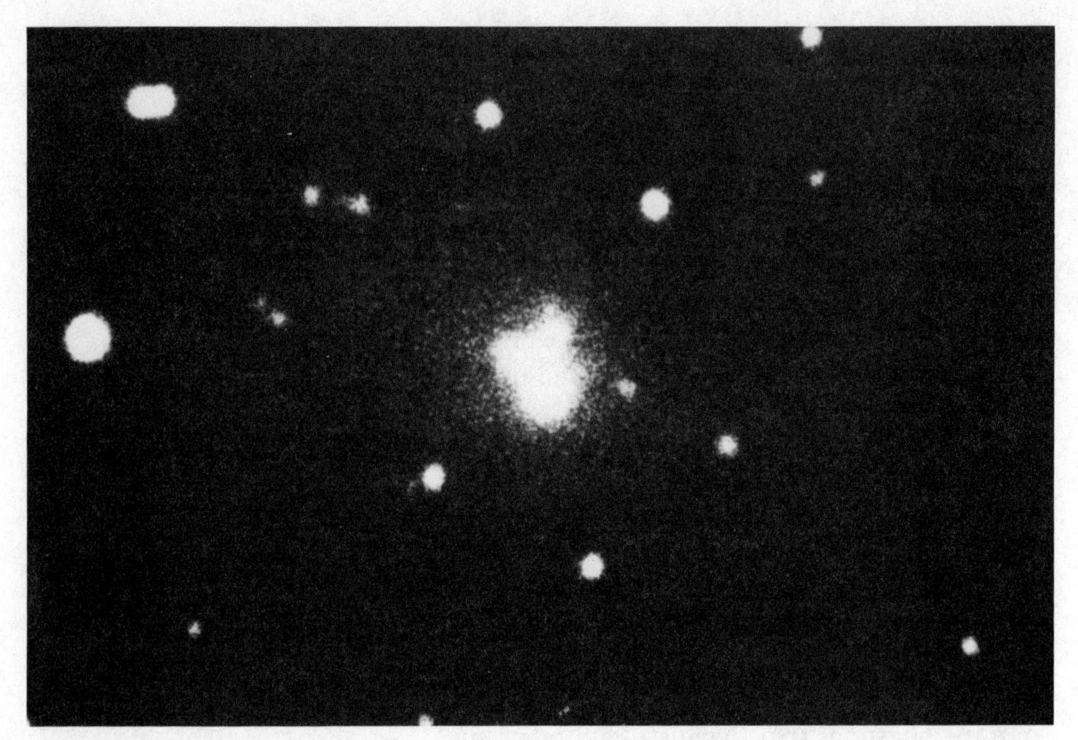

Photograph of the parent galaxy of the DRS Cygnus A. (Courtesy Palomar Observatory.)

This shortcut is called *interferometry.* It takes advantage of the key to resolving power—the diameter of the dish. Instead of using one gigantic dish, astronomers now rely on several smaller dishes that are spread out over a wide area and then connected electronically. The synthesized array has the same resolving power of a dish whose size equals the largest separation between dishes. So at a fraction of the cost of building impossibly large dishes, an interferometer gives the same resolving power.

Radio interferometry is a neat idea that translates into a complex problem in electronics. Each antenna in the array must be moved to accommodate the rotating sky, and the outputs of each dish must be consistently recorded, controlled, and stored. To make these chores feasible, scientists used electronic computers, noted for their great speed and capacity for storing large amounts of information.

Sir Martin Ryle, a British engineer and astronomer, was quick to exploit the prospects of interferometry in viewing the DRS's. He wanted to get the best radio pictures possible at the lowest cost, but this meant connecting hundreds of separate dishes. He had great insight, however. By connecting a few antennae and then observing the same DRS for 12 hours or more, he could let the rotating Earth change the position of his antennae relative to the sky, thus eliminating the need for dozens of more dishes. Using this technique, called *Earth rotation aperture synthesis,* Ryle obtained not only unparalleled radio pictures of the DRS's but also the Nobel Prize in 1974.

By the late 1960s, interferometers, employed using Ryle's techniques, were giving some odd results on the DRS's. First, most of the radio sources in the sky turned out to have the DRS structure. Second, DRS's were showing up in quasars, virtually identical in structure to those observed in galaxies; quasars looked very different from galaxies at visible wavelengths but were very similar in appearance in radio pictures.

But the most striking new result was the majestic scale of these radio sources. Obtained through the finding of the parent galaxy or quasar's red-

Cygnus A, the archtypical DRS, as shown by a radio photograph synthesized with the VLA. The visible galaxy would be a point at the middle of this scale. (Courtesy National Radio Astronomy Observatory.)

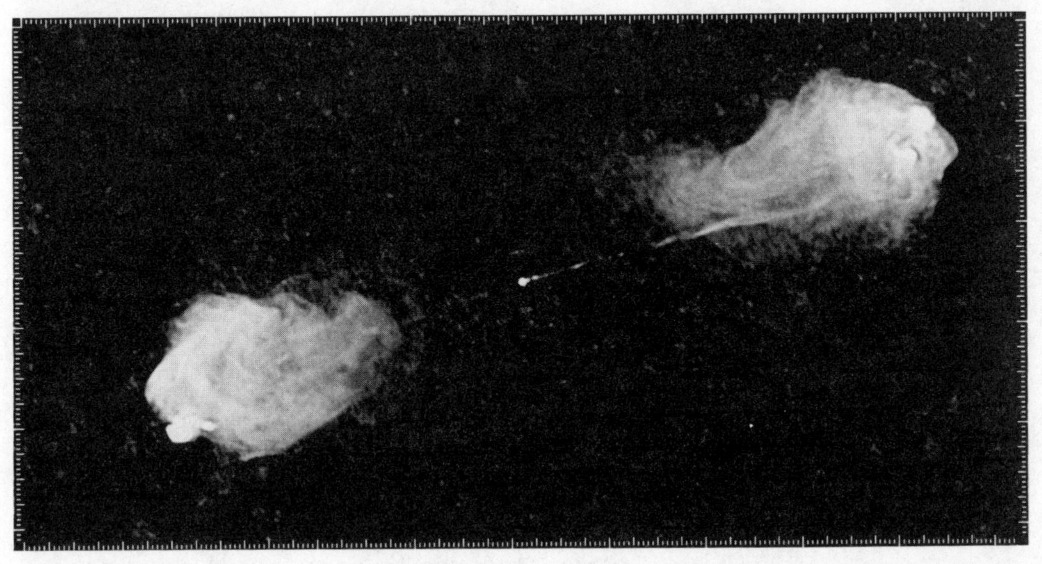

shift, the distance and angular size of some DRS's revealed that they were more than two megaparsecs across! This was hundreds of times bigger than the visible parent galaxies' or quasars' extents. Clearly, visible light views of galaxies and quasars showed only a fraction of the overall structure of these as objects. But how could spiraling electrons stray so far and exist the millions of years it took to get so far away? And what caused and controlled the huge magnetic field that must be present?

Among the latest generation of interferometers is an instrument called the Very Large Array (VLA). A wonder of metal, electronics, and computers, the VLA is spread out over a huge valley in New Mexico (larger than some island countries), which it shares with cattle, jackrabbits, and tumbleweeds. (See Plate 3 following page 78.) It is a state-of-the-art engineering feat and the interferometer that has given the greatest insight into the DRS's.

With its high resolving power, the VLA has revealed the structure of hundreds of DRS's, leading to a general formulation of their structure and the first compelling hints on the DRS process. First, at the core of the parent galaxy or quasar lies a faint and minute radio source. From this, narrow "jets" emerge, extending for hundreds, and often thousands, of kiloparsecs before they dissolve into the much larger, puffy lobes of the DRS. The lobes seem poorly shaped and turbulent, much like two radio emitting clouds. Occasionally, each lobe shows an edge, as if it were hitting an invisible barrier. As a comparison, imagine a smokestack billowing steam into the air at high pressure. At first the steam rushes out as a narrow pillar. But farther away from the stack, the steam slows down as it comes in contact with the air and forms larger, turbulent eddies before eventually dissipating. The steam has been shot out as a jetlike plume, only to be broken down by the cooler air.

In a crude sense, the structure of a DRS is similar. Although findings are not conclusive, astronomers believe that something at the heart of each DRS acts as an engine that shoots out hot gas and also makes a strong magnetic field. The magnetic field acts like a pipe, funneling the electron-rich gas away from the engine in the form of two narrow jets. Eventually, the magnetic field dissipates, and the rapidly moving jets of gas are slowed down by a tenuous but all-present cooler gas that surrounds the galaxy. The slowing jets form a barrier when they come upon the cooler gas, and vast eddies of hot gas bunch up into the DRS lobes. We see only the radiating electrons that make up the hotter gas. The engine may be spewing out the hot gas for a long period of time (millions of years), and the jets may be moving the hot gas at speeds very close to the speed of light. As a result, the lobes are almost continually supplied with more hot gas, so that as the older material cools and dissipates, it is continually replaced.

It would be wrong to let the word gas suggest that airlike densities are involved. In fact, the cooler gas surrounding the galaxy (called the intergalactic medium) is so sparse that it is thinner than the most-perfect vacuum achieved in the laboratory. The density of the hot gas isn't much higher, either. The intensity of the radiation and the size

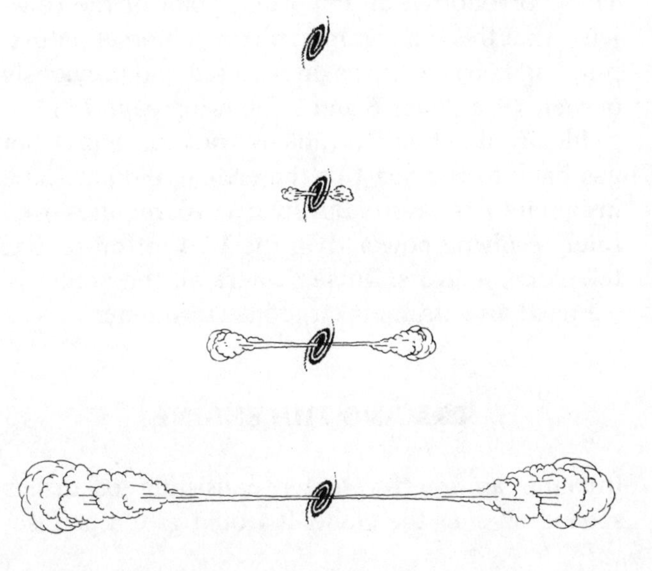

Idealized schematic of the growth of a DRS, compiled from DRS radio photographs. The engine at the core of a galaxy starts to squirt out hot electrons along two jets. Over millions of years, the jets expand farther out into space, eventually forming the two DRS lobes. Continued jet activity continually forces material into the lobes.

of the lobes allow us to make an estimate of the hot gas's density, while the distance that it takes to stop the jets betrays the density of the cooler gas. Calculations reveal that for all their splendor and breadth, the DRS lobes contain an amount of matter that is only a fraction of that in the parent galaxy, and the intergalactic medium has a density that is just a small fraction of the density needed to close the universe—ρ_{CRIT}. Even though the DRS's may be considered the largest objects in the universe, they are mere shadows of mass when compared with the visible galaxies. And if there is any great amount of dark matter in the universe, it is not contributing directly to the makeup of the DRS's. (See Plates 4 and 9 following page 78.)

Not long after the first radio jets were seen with the DRS's, very sensitive photographs and visible-light electronic images revealed that many jets also gave off dim but perceptible visible light. In fact, some galaxies had visible-light jets where none could be seen in the radio. These visible jets are caused by the same engine as the radio jets, although perhaps they are the result of electrons moving at higher energies. The visible jets form an eerie counterpoint to the usual pinwheel swirls and fuzzy star clusters of their parent galaxies. They corroborate an important point of the radio jets—that this is a common phenomenon of nature, puzzling because it was unexpected and previously unseen. (See Plates 6 and 7 following page 78.)

Ideally, we should be able to trace the jets of hot gas back to the heart of the engine and get some insight into its nature. But such a task requires even finer resolving power than the VLA, or an optical telescope, delivers. To see where all the action is, we need an extremely large interferometer.

TRACING THE ENGINE

Imagine an interferometer consisting of dishes spaced all over the globe: It would give a precise-

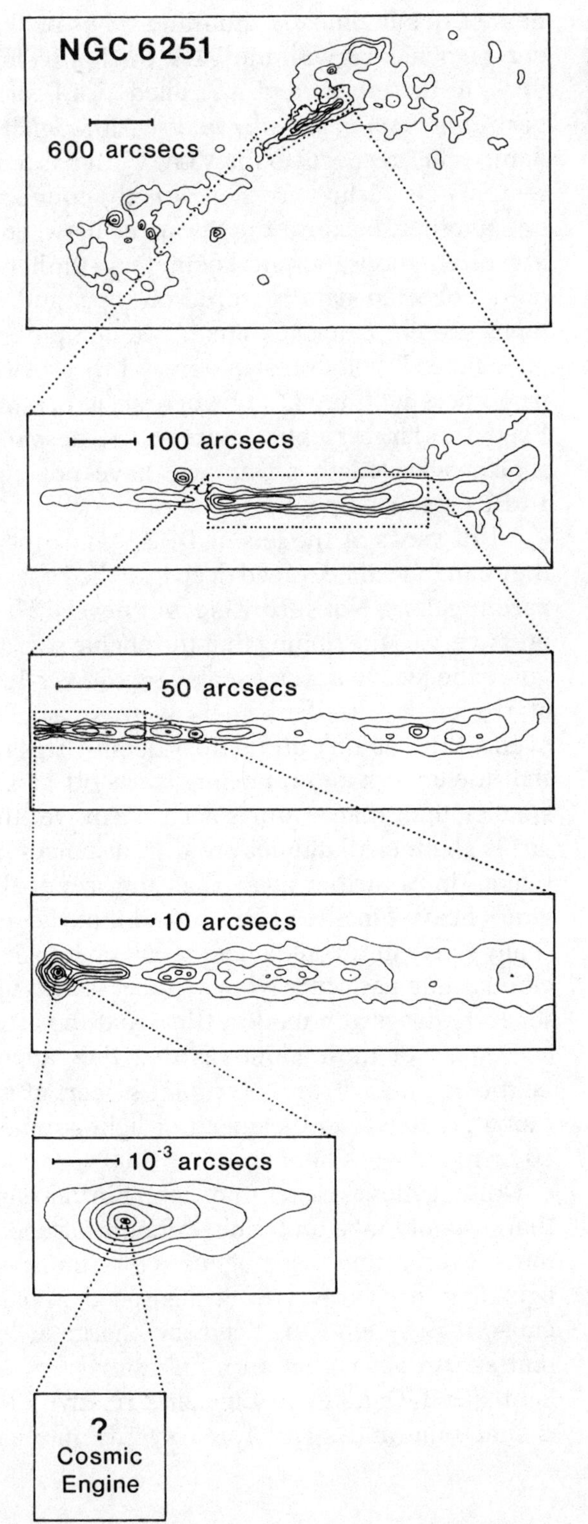

Composite drawing of the same jets as in Plate 4 (following page 78), but showing how VLBI probes the core and inner jet. (Courtesy Alan Bridle, NRAO.)

ness of detail almost a thousand times better than can be achieved with the VLA. Such a technology does, in fact, exist, and it's called VLBI, short for *very long baseline interferometry.* Although it's similar in other respects to the VLA, VLBI has a unique feature—its dishes are not directly connected to one another, because the distances between them are often thousands of kilometers. Rather, each one's collected data is stored on tape, and all the tapes are electronically linked via computers at a later time. Through this system, VLBI provides astronomers with images of unparalleled resolution. Even the largest optical telescopes, despite their shorter-wavelength advantage, have not equaled VLBI's resolving power.

VLBI views of the jets of DRS's confirmed that they can indeed be traced deep into the heart of the parent galaxy. Not surprising, but nevertheless impressive, was the finding that the engine was able to guide the jets over a size scale of well over 100,000; at the engine itself, the jets were so small that even VLBI couldn't probe their details. This means that the energy and funneling starts off in a space smaller than that of our solar system, yet the hot gas is channeled and delivered to distances over a billion times farther away than the size of the engine's heart. Finally, VLBI showed that the jets actually start out turbulent and break up into discrete clouds, much as water from a faucet breaks up into separate drops at a modest flow. And the astounding aspect of these clouds is that they appear to be moving away from the engine's heart at speeds much greater than the speed of light—sometimes 20 or more times light speed.

Nothing moves faster than light. Einstein showed that it would take an infinite amount of energy to move even a tiny piece of matter to the speed of light; light itself traverses at this speed, in part, because it is weightless. Yet very energetic bits of matter can get within an infinitesimal fraction of light speed. One key to Einstein's relativity theory is that time and space for a swiftly moving ob-

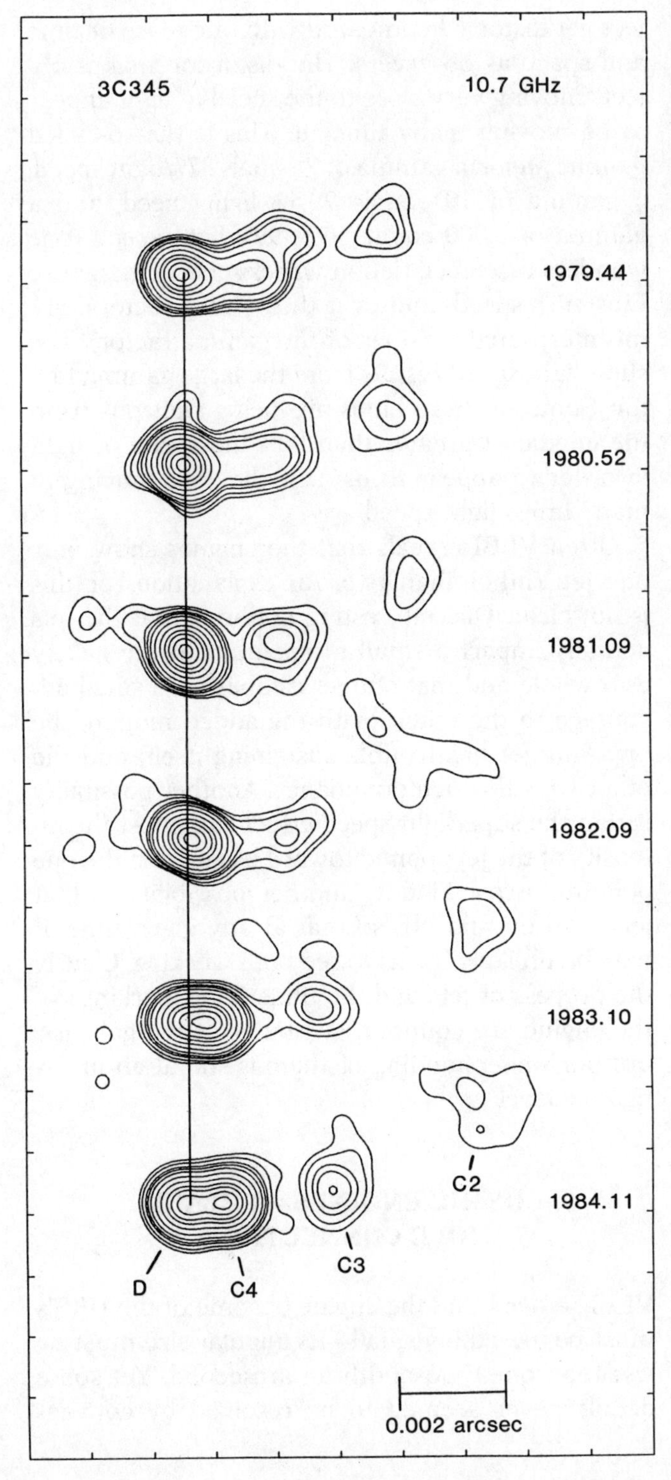

Quasar 3C345 is one of the DRS's that shows faster-than-light motion in its inner jets. Here, clouds of electron gas (marked accordingly) have been shot out and show a change of position over the course of years. (Numbers at right indicate time of observation.) The angular change, though only thousandths of an arcsecond per year, translates to incredible speeds at the distance inferred from the redshift of the quasar. (Courtesy John Biretta.)

ject get distorted compared with our sense of time and space as observers. The distortion makes objects moving very close to the speed of light appear to be moving many times it. This is the so-called *gamma factor;* a gamma of 2 equals 87% light speed, a gamma of 10 equals 99.5% light speed, and a gamma of 1,000 equals 99.9999% light speed (one wonders if science fiction writers choose to ignore Einstein's speed limit or if their "warp factors" are misinterpreted versions of the gamma factor). The superlight speed results from the large gamma factor. Some of these blobs are being shot out from the engine at greater than 90% the speed of light in order to appear to us as if they are moving at many times light speed.

Often VLBI reveals that the engines show only one jet, rather than two. An explanation for this is not clear. One interesting notion is that the jets actually impart a small momentum to the galaxy as a whole and that one jet establishes a small advantage to the other. With the added motion, the opposing jet has trouble sustaining itself, and the other wins through dominance. Another possibility is that the superlight-speed effect can boost the intensity of the jet pointed toward us and dim the one pointing away. Finally, another possibility is that jets turn on and off, so that at any given time, it may be unlikely for us to see both working. Clearly the process of jets and the underlying workings of the engine are common attributes of the galaxies, but our understanding of them is still at an investigative level.

COSMIC ENGINES—A BLACK HOLE CONNECTION?

VLBI showed that the engine of some of the DRS's must be incredibly small—its angular size must be less than one-thousandth an arcsecond. Yet some details never seemed to be resolved by conventional VLBI. In 1987, however, German astronomer Norbert Bartel and his colleagues undertook a special VLBI observation. It was unique in that it used very short radio wavelengths, much shorter than those previously used, to get superhigh-resolution radio pictures of the core of a radio galaxy, dubbed 3C84. For the first time, this observation revealed a radio region less than a tenth the size seen in previous radio observations. The region looks not unlike a sphere surrounded by an extended pancake-shaped ring. And for so much radio emission to be coming from such a small region precluded the likelihood of it being a collection of many stars; rather, it suggested one compact star at least a million times more massive than the sun. This finding strongly pointed to a black hole as the engine.

A *black hole* is a collapsed star that, perhaps in the process of becoming a supernova, imploded upon itself. Its mass may remain about the same, but its diameter has shrunk to an infinitesimal fraction of its former state. This peculiar situation leads to a bizarre condition: The light emitted from a black hole never leaves it. It is as if the light is drawn back by the magnified gravitational field caused by this shrinking. This is no ordinary star—densities are so high that even neutrons and electrons cannot exist on its surface but are torn apart by the pressure of collapse. A black hole is, effectively, a point in space with the mass of a normal star or more.

The problem with having so much mass in a small area, as the 3C84 observation implies, is that the gravitational attraction of the mass upon itself will force the matter to pull together and collapse into a black hole. It is impossible to maintain a normal star with these conditions.

In 1970, British physicist Martin Rees modeled the role of a black hole in the DRS's. He discovered that a black hole can act as an engine, drawing in matter around it, then violently pushing it out

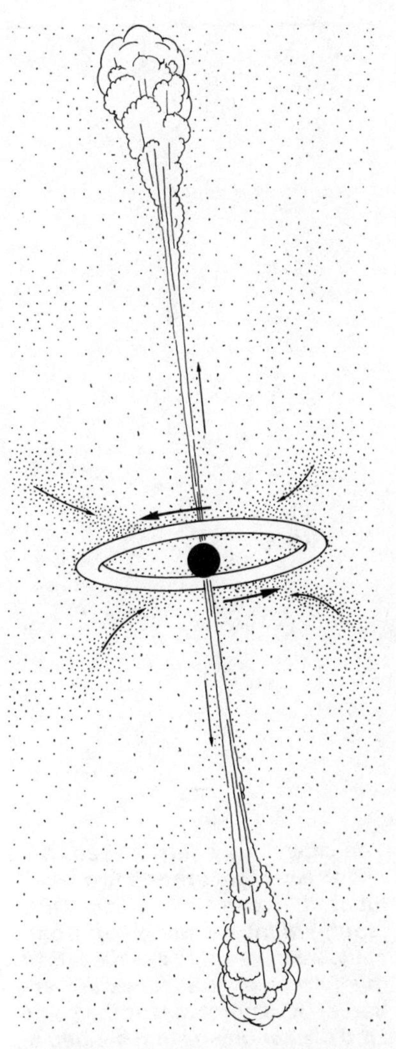

A black hole engine can draw in gas, forcing it to collect in a rotating, pancake-shaped accretion disk. Some of the gas falls into the black hole, while other gas is superheated and shot out along the poles of the accretion disk (perpendicular to the disk). The hot gas forms the two opposing jets.

again in funnels formed by the black hole's magnetic field. Before the matter is drawn in, it forms an *accretion disk,* a donut-shaped ring surrounding the black hole, that itself emits radiation. The disk is a conflagration of superheated gas in the most violent of states. The 3C84 observation bears an uncanny resemblance to the prediction made by Rees' theory. VLBI and VLA observations also have been providing better hints on the workings of the DRS engine through studies of objects much closer to home. These objects are *jetstars,* two-star systems whose study reveals very compact stars (although a black hole need not be assumed) that draw out gas from much larger (in size, not mass) companions, forming a pancake-like disk. As the gas from the disk attempts to fall onto the compact star, the gas is heated up and ejected, and it gets squirted out the poles forming the two jets. The expulsion of the jet gas eventually forms DRS-like lobes, although the lobes may start out being quite small and the jets themselves may be very dim. Only two jetstars have been identified in our galaxy so far—they are called SS433 and Sco X-1. Others are suspected

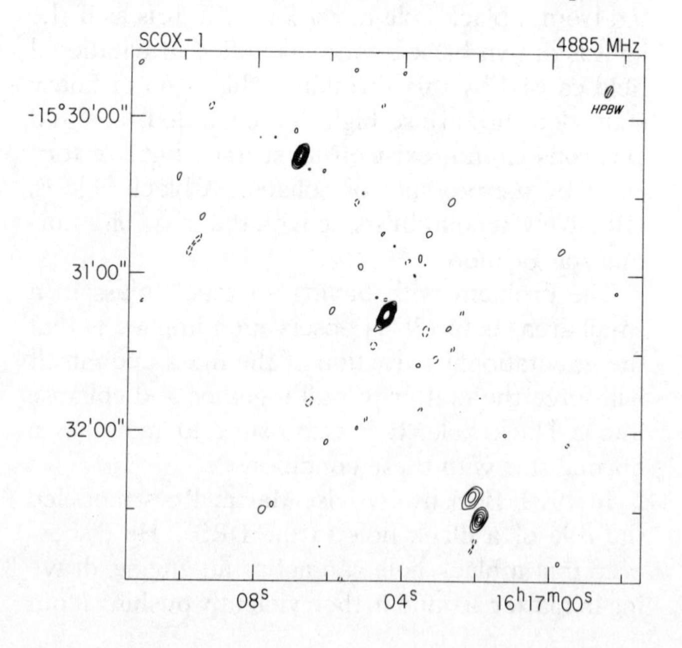

The Milky Way jetstar Sco X-1 shows a center core and two small lobes that one gets with a DRS (southern lobe is doubled). Ironically, we have yet to see the jets in this jetstar; only its lobes are apparent. It may represent the same kind of phenomena that happen at a DRS's engine. (Courtesy Barry Geldzahler.)

and await further observations. If the engines in galaxies and quasars are much more massive versions of these jetstars, then it may be easy to study the jetstars over time to find out the details of how the gas gets squirted out, how the magnetic fields channel the gas, and what makes the jets break down into lobes.

SPECTRAL-LINE RADIO ASTRONOMY

From its inception, radio astronomy sought the development of even more sophisticated equipment and techniques to probe the DRS's. But another important aspect of radio astronomy is unrelated to the DRS's and has much in common with a technique of visible-light astronomy.

Recall that a prism can be used to break up light into a spectrum of narrow wavelengths. This reveals the spectral lines of molecules and atoms, which can themselves be used to find velocities through the Doppler effect. But not all spectral lines happen at visible-light wavelengths. Some of the most salient ones occur in the radio region of the spectrum. *Radio spectroscopy* can be used as a companion tool to optical spectroscopy to corroborate or elaborate conditions of stars and galaxies.

In radio spectroscopy, the astronomer observes a range of frequency channels and records the relative intensity in each; the process is similar to the one found with police scanner radios. The kinds of spectral lines sought are ones from common components of the universe—cold gas (again, in very thin densities) from hydrogen, water, carbon molecules, and others. Some of the radio spectral lines give clues on the formation and growth of baby stars, while others display wisps of matter in what appears to be otherwise empty space.

By far the most important such spectral line has been the 21-centimeter hydrogen line (21 centimeters refers to the wavelength of the spectral line). By observing this line, vast regions of cold, tenuous hydrogen atoms can be found, filling in some spaces between stars, surrounding the environments of clusters, and tracing the motions of the turning of galaxies. As much as 10% of the entire matter of a galaxy may be this pervasive hydrogen gas, easily shown through the 21-centimeter spectral line. A useful feature of this 21-centimeter spectral line is that it reveals, through the Doppler effect, how different points in a galaxy have different velocities, which are caused by the galaxy's rotation. In effect, the cold hydrogen gas orbits the galaxy's core, and the slight red- and blueshifting of the 21-centimeter spectral line indicates how this orbiting occurs. The 21-centimeter line shows that areas close to the core move at much faster velocities than does the gas much farther out.

One extraordinarily useful aspect of all of this is the overall range of velocities of the 21-centimeter line from a galaxy. It may amount to several hundreds of kilometers or more per second. In the mid-1970s, American astronomers Brent Tully and Robert Fisher discovered that the absolute magnitude of the core of a galaxy was closely related to the range of the 21-centimeter hydrogen-line velocities, such that the brighter galaxies had greater velocity ranges. The reason for this is not well understood yet, but it is believed that the larger velocity range indicates a greater galactic mass, which also tends to give a greater brightness. The beauty of this relationship, now called the *Tully–Fisher relationship,* is that it provides a totally new way of estimating absolute magnitude for a galaxy, quite independent of other techniques, such as Cepheids. Recall that absolute magnitude, combined with the measured magnitude, can be plugged into the distance modulus to get distance. The Tully–Fisher relation is another yardstick for cosmic distance measurement, made possible through radio spectroscopy.

A useful spin-off of this new yardstick is that it

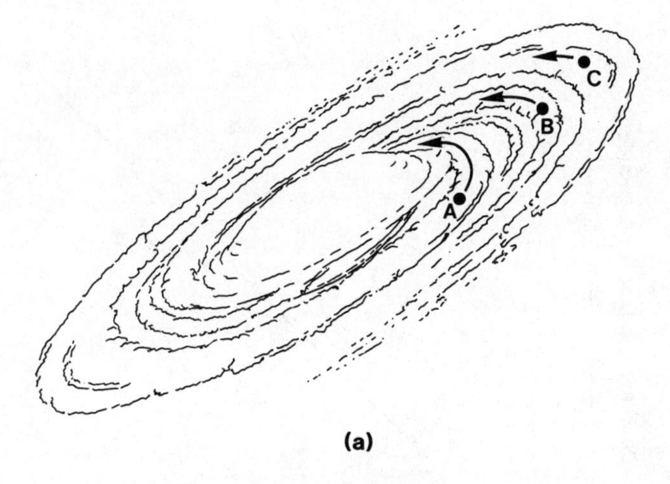

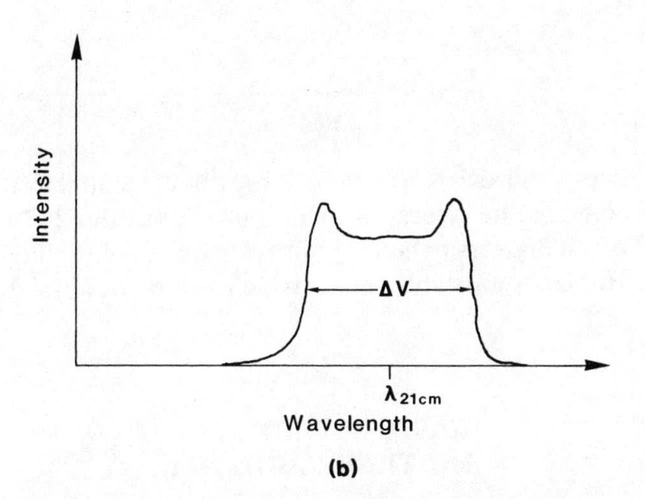

Different points along the arms of a spiral galaxy (a) move at different speeds; the closer to the core, the faster the motion (in general). Looking at the 21-centimeter hydrogen line (at the redshift of the galaxy), we see not a line but a broad profile representing the overall velocity range, Δv, of the galaxy's rotation (b).

allows a way of calibrating the Hubble diagram in finding H_0. This is accomplished by assuming that the distance follows Hubble's law, then using the redshifts of the galaxies to derive Hubble's constant. Although this method is but one of many ways to derive Hubble's constant, it shows particular promise and is a unique contribution that radio astronomy makes to this key estimate of a cosmological parameter. Using this approach, the American astronomer Allan Sandage and the Swiss astronomer Gustave Tamman have obtained a value of $H_0 = 55 \pm 15$ km/s/megaparsec. Some contro-

J. R. Fisher and R. B. Tully found a relation between the velocity range, Δv, of a galaxy as seen with the 21-centimeter line and its absolute (infrared) magnitude (a larger negative value of absolute magnitude indicates greater intrinsic brightness). This information can be used to derive a value of Hubble's constant through distance calibration.

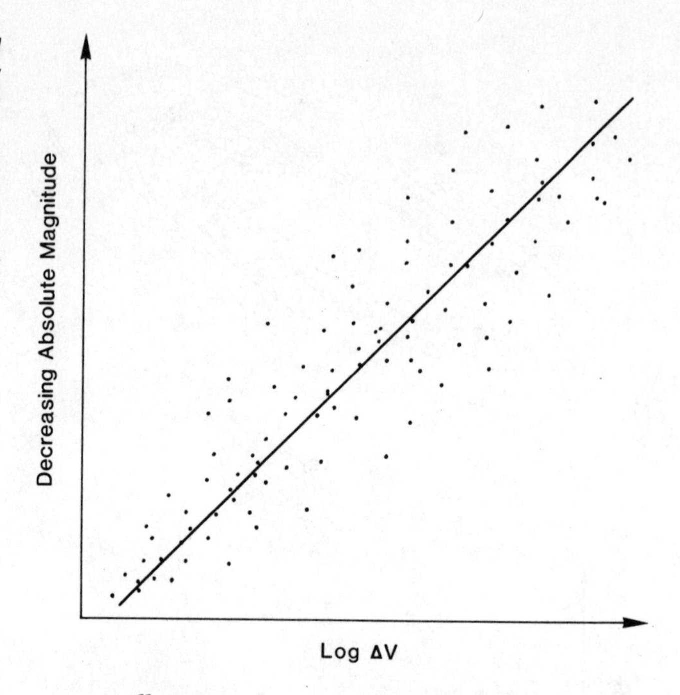

versy still exists about whether they account for some subtle effects in their analysis, but this 1979 result appears to hold up with other ways of finding Hubble's constant, one of which will be described next.

RADIO SUPERNOVA—ANOTHER YARDSTICK

Not all the radio sources in the sky are DRS's; occasionally a star will be found to emit the synchrotron radio emission. One extreme case of this appears with supernovae, the brilliant and violent deaths of some stars. During some stages of a supernova explosion, the explosion pushes out a shell of gas that gives off a radio emission. This same gas might also emit a visible-light spectra, from which it is possible to use the Doppler effect to get the redshift of the supernova, as well as the velocity by which the shell expands. (See Plate 11 following page 78.)

Astronomers have long known that distance can

be inferred by looking at an expanding shell—if you know the velocity of expansion. For example, if a shell appears to be an arcsecond in size one day and two arcseconds the next, then the supernova must be at a distance where one arcsecond corresponds to the increase in physical size. The expansion velocity multiplied by the time between measurements equals this size; with angle and size known, it is then possible to solve for the distance. No supernova is close enough to give these kinds of angular sizes. All supernovae are so distant that

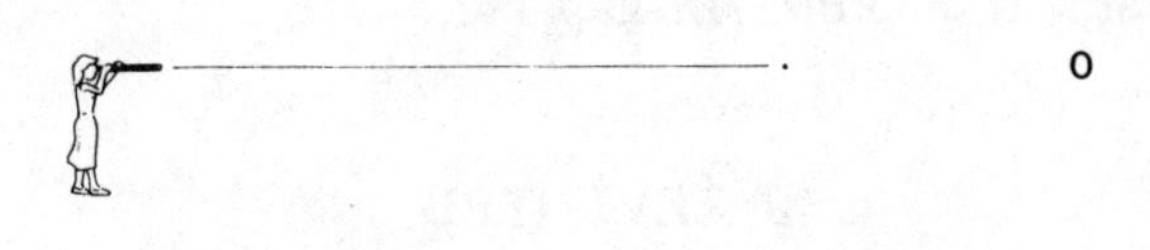

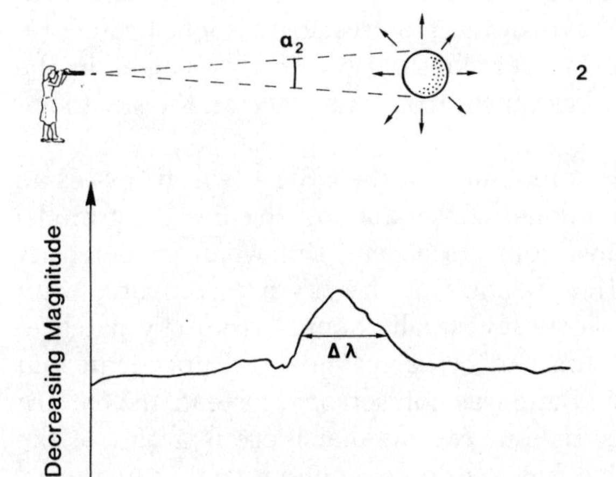

An exploding supernova becomes an expanding radio source whose angular-size change can be measured with VLBI. The spectra seen with visible light (bottom) shows a velocity width from expansion. With the expansion velocity and angular-size change known, it is possible to infer the distance.

their angular sizes are measured in thousandths of an arcsecond—an angular-size scale that can be probed only by VLBI.

That is exactly what Norbert Bartel has done. Using VLBI, he has shown that the angular size of the radio emission from a supernova in the galaxy M100 increased after several months. Using this data to solve for the distance, he then found Hubble's constant. The result is not unlike that derived by Sandage and Tamman: $H_0 = 65 \pm 30$ km/s/megaparsec. Clearly we are getting closer to finding this cosmological parameter. There is much confidence in the idea that H_0 lies at or near the value of 60 km/s/megaparsec.

CBR REVISITED

As described in Chapter 2, the discovery of the CBR is credited to radio astronomy. CBR studies are a very different aspect of contemporary radio astronomy because the CBR is not like the DRS's, for example. Rather, the CBR's continuous glow challenges us to study it using something other than interferometry or spectroscopy.

Although radio astronomers have not solved all the problems raised by the discovery of the CBR, they have devised increasingly sophisticated receivers to search for very small changes in the CBR's brightness from one place on the sky to another.

The smoothness of the CBR—its isotropy—is an observational cornerstone of the Big Bang model we have come to accept. But when the intensity of different parts of the sky are compared with great accuracy, small changes from sky patch to sky patch can be seen. This does not mean that the Big Bang was not isotropic; instead, the pattern of fluctuations reveals that there is a part of the sky that has a little more intensity than the average, while the area in the opposite direction is less intense by the same amount.

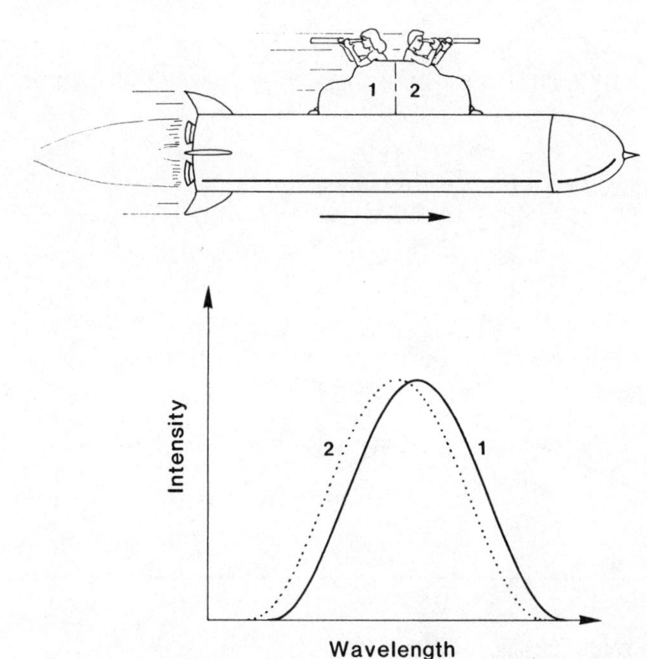

Two observers of the CBR look toward and away from their direction of motion and see slightly different intensities of the CBR; this is one way of getting the anisotropy, here a fractional difference due to the Doppler effect with respect to the CBR by the galaxy's motion.

This effect, first discovered by the American physicist Brian Corey in 1977, might best be understood through the Doppler effect, in which our galaxy has an extra bit of motion from the general expansion of the universe. Here, the Doppler shift will cause the "preferential" direction in which we are heading to be slightly more intense and to be slightly less so in the opposite direction. The question remains: Why do we have any preferential direction at all? Is something drawing our galaxy toward it, imparting a velocity that has nothing to do with the expansion of the universe?

COSMOLOGICAL TESTS

As we have seen earlier, cosmological tests are ways of comparing observations to models to seek out the values of the cosmological parameters. The Tully–Fisher relation is actually a cousin of the Hubble law and is considered an important contribution of radio astronomy to cosmology. The cosmological tests described in Chapter 3, however,

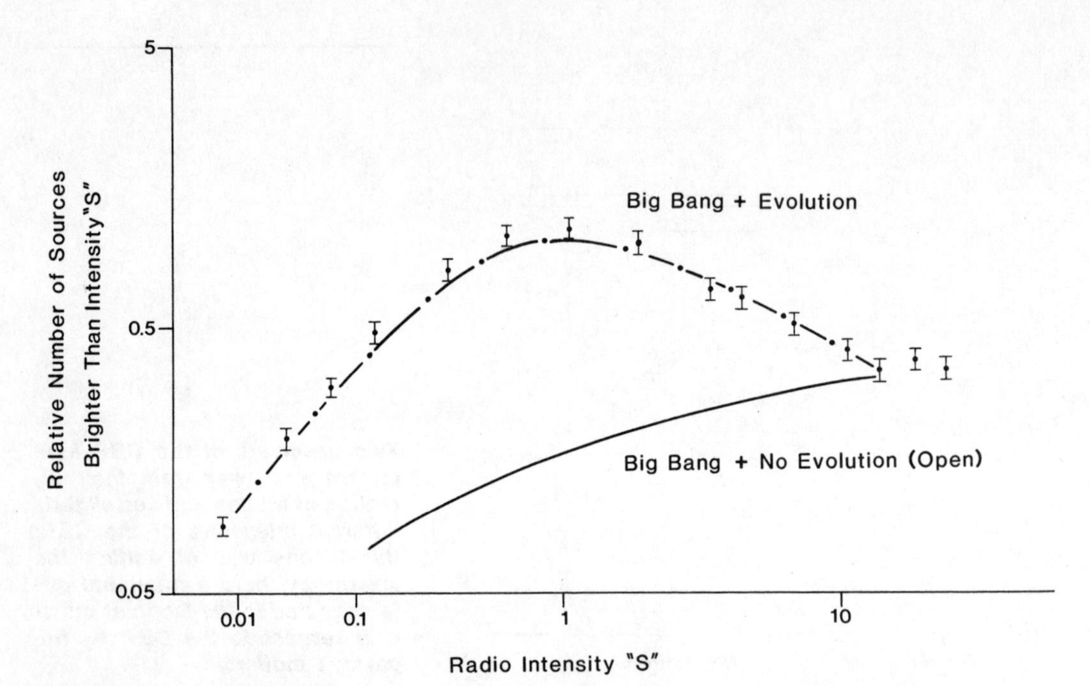

The radio source counts test, showing how evolution affects the models (dotted line).

also have some radio counterparts. Although these are still limited in their success, they are an ongoing part of radio astronomical studies. Here we describe these radio counterparts.

By far the most useful of the cosmological tests is the source counts test. Here, the sky is searched for radio sources (such as DRS's), and tables are made of the number of sources in increments of decreasing intensity values. A simple model for the effect of galactic evolution (change in radio brightness of galaxies over time) can be devised. Coupling this with the expected effect of the Big Bang, it is seen that the source counts are consistent with the Big Bang model, corroborating, but by no means providing unique proof of, the model.

The radio angular-size test is also similar to the one using visible light. The principle is as follows: DRS's have angular sizes, which, if they correspond to the same physical size, should appear to grow smaller and smaller with increasing redshift. But at some very large redshifts, the angular sizes may start to look bigger because of the curvature of the

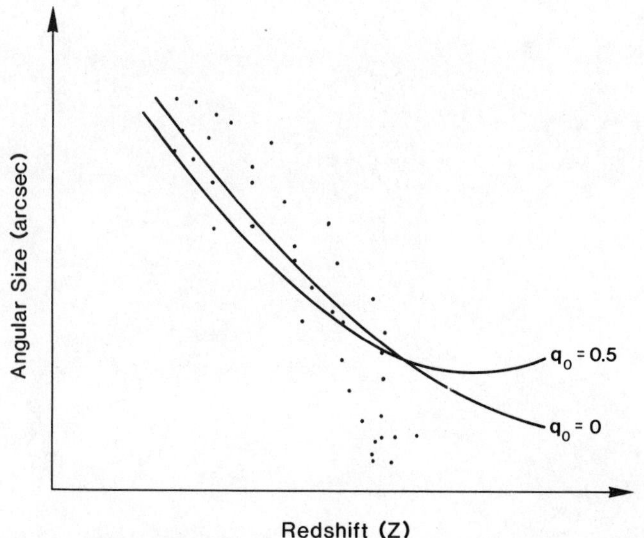

The radio angular size-redshift test for DRS's. DRS's actually look smaller than expected from the universe's curvature; evolution may be the culprit.

universe. Thus a plot of angular size versus redshift may reveal the curvature of the universe.

The angular-size test shows not only that DRS's do not have larger angular sizes at large redshifts but also that they actually are smaller in size at large redshifts! This could be a galactic evolution effect, where the higher density of the intergalactic medium forces DRS jets to form the lobes closer to the parent galaxy. In any case, the decrease in size is especially complex from the view of almost any cosmological model that excludes galactic evolution; the Big Bang is not especially at risk here.

But, once again, galaxies are entwined with cosmology. We need to see how better understanding of them has progressed in the last decade.

· FIVE ·

Galaxies, Quasars, and the Evolution Revolution

GALAXIES are markers for the expanding universe, keys to finding out how the universe changes as it expands. But as the problem of galactic evolution makes clear, the galaxies are not necessarily a collection of identical markers. They have their own variety of differences independent of those brought about by the expansion of the universe. So to understand the tale of the universe means understanding the galaxies as well; what makes a galaxy what it is, and why does it change?

TYPES OF GALAXIES

There are billions of galaxies in the universe, but our unaided eye can see only a handful besides our own Milky Way. Even with the aid of high-powered telescopes, the galaxies appear dim, some extraordinarily so. This presents a special problem in studying their various properties and in classifying them by these properties.

In general, every galaxy has a bright, compact central region, or *core.* Most of the stars in a galaxy are located in or near this core. Outside the core are extended or elongated areas of stars, and often gas and dust, which in photographs do not show up as brightly as the core. We have been able to identify particular shapes and structures of galaxies that are bright enough to be seen with telescopes. Some appear as fuzzy blotches, while others have a distinct spiral formation with radiating arms. A galaxy's image, obviously, is one of the main criteria we use to classify it, along with such other

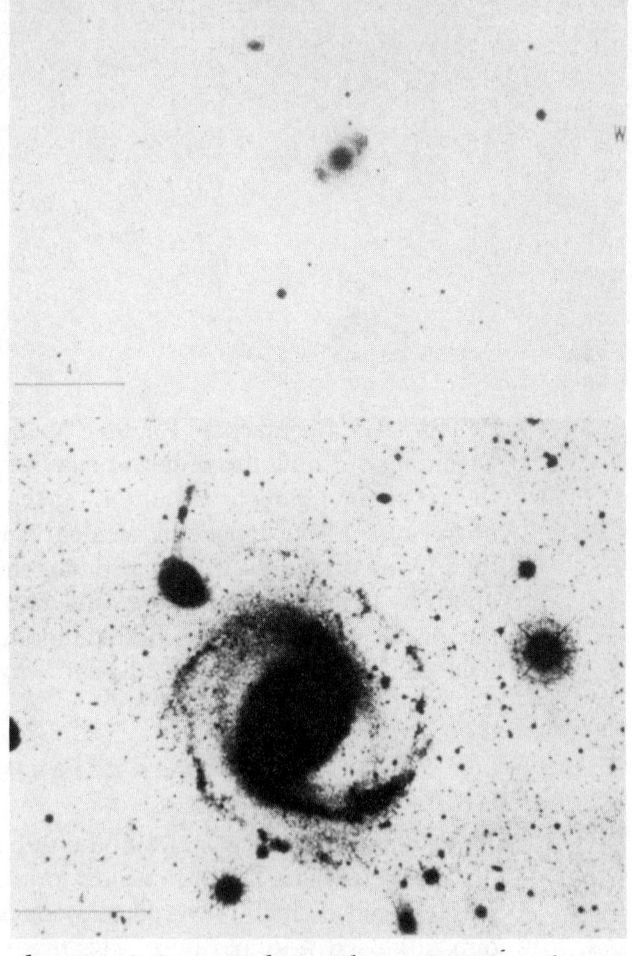

Two views of an active galaxy taken with two different telescopes of vastly different sizes. Only the core appears on the smaller telescope's exposure, while the spiral arms become apparent with the larger telescope's more sensitive exposure. (Courtesy H. C. Arp.)

characteristics as color, radio emission, and spectral lines. But the most dramatic feature that distinguishes the galaxies from one another is how their cores brighten or dim. If, like our own galaxy, the core does not change appreciably with time, the galaxy is called a *normal* galaxy; otherwise, it is called an *active* one. (See Plates 12 and 14 following page 78.)

Normal Galaxies

Most of the galaxies that have been observed with telescopes are normal galaxies. They come in many

An elliptical galaxy. (Courtesy Caltech.)

different forms, although they can be divided into three major groups, based on shape and structure.

In his graduate thesis, Edwin Hubble proposed a classification scheme for the normal galaxies, and his scheme is used to this day. Hubble had in mind more than classification—he suggested that one type of galaxy evolved into another over time. (Curiously, the idea of galactic evolution did not escape Hubble's attention. He believed that evolution would apply mostly to shapes, not absolute magnitudes.) Although no one has shown that this evo-

A galaxy's orientation angle can make it look very different. Here a spiral galaxy is edge on to us. (Courtesy Palomar Observatory.)

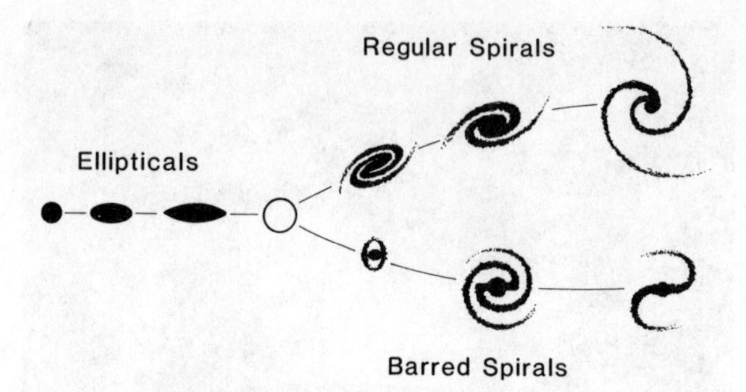

Hubble's "tuning fork" diagram, a classification scheme for the normal galaxies.

lutionary chain goes as Hubble described, it does offer the convenience of distinguishing one class of galaxy from another.

A logical way to distinguish galaxies might be by their mass; however, with many extreme cases, galaxies show a large range of masses. Instead, Hubble divided normal galaxies into three major groups based on their appearance: the spirals, the barred-spirals, and the ellipticals. Not all normal galaxies fall into these three categories (some others are called irregulars, dwarfs, and peculiars), but a majority do. *Spirals* are pinwheel-shaped galaxies, with several arms projecting from the core. *Barred spirals* have only two or four arms, which are connected to the core through a barlike structure. *Ellipticals* are cigar-shaped, or spherical, star groupings, lacking spiral arms. Many look like globular clusters save for their much greater number of stars and redshifts.

There are two big differences between the latter type and the first two, differences that account for at least some of their outward distinction. The first is *angular momentum* (not related to angular size), a term from physics that describes the speed and direction with which stars move about, or orbit, the galactic core. Spectral-line studies, such as those done at radio wavelengths with the 21-centimeter hydrogen line (see Chapter 4), reveal that the spiral arms rotate (an entire turn occurs every 100 million years or so). This rotation, which is slightly different for stars farther out from the core, gives the arms their twirled appearance in barred and nonbarred spiral galaxies. In contrast, the outer stars of elliptical galaxies show little, if any, rotation. The elliptical galaxies have little angular momentum, while the spirals have a lot.

The other major difference is the gas content of these galaxies. The 21-centimeter spectral-line studies show that elliptical galaxies contain little gas—at most, a few percent of their total mass. Spiral galaxies tend to contain a significant amount of gas, perhaps 10% or more of their mass; this gas, coupled with dust, is what gives the spiral arms a mottled appearance. And because new stars form out of gas, spiral galaxies are said to be centers for star formation, while ellipticals, in general, contain very old stars and show little, if any, new star formation. The arms of spiral galaxies also appear more blue than their cores, indicating that these sites of star formation contain young, hot stars.

The Active Galaxies

Normal galaxies are differentiated by their shape, but the active galaxies are distinguished by the radical and rapid magnitude changes in their cores. An active galaxy may resemble a normal galaxy with a very strange core, within which violent and energetic behavior can be seen. (See Plate 8 following page 78.)

As in the quasars, the flickering of magnitude in the core, over days, months, or years, indicates that stars are being torn apart by some central force. It should not seem surprising that nearly all of the prominent DRS's radiate from active galaxy parents, which undoubtedly contain cosmic engines. Spectral analysis of active galaxies indicates a large range of velocity of motion within the cores, perfectly logical if stars and gas are being sucked in or alternatively, if gas is being shot violently outward, perhaps in the jets that form the DRS's. (Black holes are likely to be responsible for the DRS activity and the accompanying rapid magnitude variations in the core.)

To appreciate the role of active galaxies, it is interesting to look at one or two with a photograph. Rarely do spiral arms appear; usually only the tiny bright core can be detected. Unlike the quasars, however, these active galaxies are not pointlike but show filaments of hot gas extending from their cores. The structure may be a little biased to our view because the brightness of the core can be hundreds of times brighter than a normal galaxy at the same distance; outer structure is much fainter for active galaxies than it is for normal ones. It is the core that makes active galaxies unusual compared with normal galaxies, but what about compared with the quasars?

QUASARS AS GALAXIES

Quasars exhibit nearly all the same properties as active galaxies. But there is a fundamental problem in understanding quasars—they lack any of the obvious extended structure seen in galaxies. It took more than twenty years to make a convincing case, but in the late 1970s several quasars were shown to be active galaxies and not something entirely different.

The problem of identification was one of sensitivity to the dimness of the extended structure of quasars. Even sensitive photographic techniques make quasars look like little more than pinpoints of light. The solution had to await the technology of the 1970s.

For more than 100 years, photography was the only way of getting images of galaxies. But the problem with film, or plates, is that even the best emulsion records only a fraction of the light that falls upon it. Fortunately, the electronics age provides astronomers with alternatives.

The most useful new way of *imaging*—the general term for recording pictures on film, computer, or otherwise—has been through a new technique that uses an array of semiconductors to record the images, absorbing almost all of the light that hits it. This technique of electronic imaging is called the *charge coupled device* (CCD) technique, and it is rapidly seeing many uses, both in and out of astronomy.

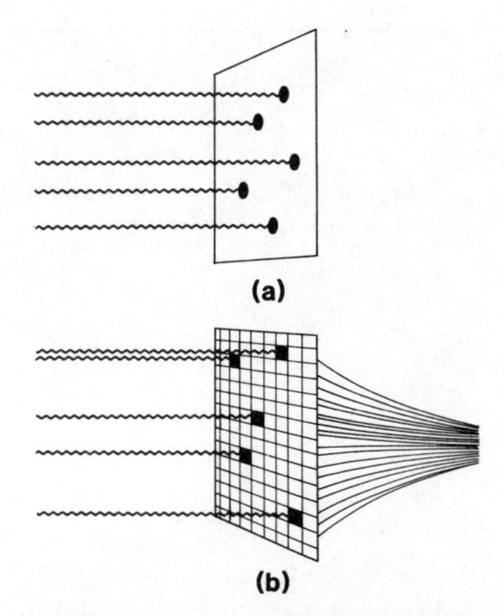

In film photography, light hits an emulsion and causes a chemical reaction that exposes the film (a). A far more efficient process occurs with the CCD method of imaging (b), where light hits individual semiconductor pixels, causing currents to flow that can be sent through wires for storage onto a computer.

With such effective recording devices as CCD's, astronomers could look at very dim light levels unobservable through photography. This was exactly the type of technique needed for the study of quasars. In 1981, American astronomer Susan Wyckoff and her colleagues used CCD's to explore one of the best-studied quasars, called 3C 273. They found a dim, extended fuzz around the bright pinpoint of 3C 273 and were able to take its spectra, which gave its redshift. This redshift matched that of the much brighter pinpoint of the quasar, clear evidence that the fuzz and the quasar were one and the same object. With such an extension, 3C 273 could now be considered an active galaxy with a very bright core and dim extended outer regions. (See Plate 23 following page 142.)

Since that time, a number of quasars have been examined for this fuzz. In most cases, the result was a reclassification of these quasars to active galaxies. Many now appear to be spiral galaxies with extraordinarily bright cores. This is surprising because it dissolves not only the large distinction between quasars and galaxies but also the one between normal and active galaxies. This suggests that what really decides if an object is a quasar is the violence of activity that goes on in the core; the most violent active galaxies are the quasars.

The identification of quasars as active galaxies demystifies them, fitting then at last into the general scheme of our galaxy-filled universe. But what makes a galaxy become a quasar, and why are quasars much more common in the distant past?

These and other enigmas may be addressed only through an understanding of how galaxies come into existence and what factors determine their fate.

HOW DO GALAXIES FORM?

Galaxies must have formed early on in the universe. We know this because we don't see anything that might look like a galaxy coming together at tiny redshifts, while at the other extreme, we don't see any galaxies or quasars beyond an inferred distance of 25,000 megaparsecs. This edge of the galactic realm was established in December 1987 with the discovery (by British astronomer Stephen Warren) of a quasar named Q0051-279 at a redshift of $Z = 4.43$. At its inferred distance, this object is receding from us at greater than 93% the speed of light and from our perspective inhabits a universe only a tenth of its present age. But the mystery remaining is how the hot hydrogen left over from the earlier time of the CBR eventually formed the stars and galaxies. I'll discuss a specific method for this process in Chapter 8, but here let's look in general terms at what astronomers now believe to have occurred.

After the period during which the CBR indicated the makeup of the universe, gravity began to exert its influence on the expanding matter. At the time of the CBR, the universe was apparently quite smooth, perhaps with few fluctuations to act as "seeds" for pulling matter together. At some stage, though, we know that small amounts of matter began to clump. And astronomers have been able to create computer simulations that mimic the process and show us how the galaxies came together.

Computer simulations show that, gathering strength from an initial seed mass, the enhanced gravity of that mass attracts, pulling in nearby, diffuse hydrogen left over from the CBR era. The seeds grow, forming neighboring eddies around it in the process. These eddies themselves become seeds, attracting their own diffuse matter and forming it into huge clumps—the earliest galaxies. The clumps, if close enough together, can tear off parts of each other as they move. If they touch, they might collide and form bigger clumps. In any case, the new protogalaxies—the embryonic form of today's galaxies—tend to stay together, forming groups called clusters.

Galaxies did not form in isolation from one another, since the formation of one influences the actions of another through the pull of gravity. The gravity from the seed galaxy attracts all the mass around it and causes it to form galaxies. Computer simulations make this early process clear, but they do not show quite so clearly how gravity continued to exert its influence long after the first protogalaxies were formed. Still, simulations are useful tools as predictors of what might be seen.

Over time, the ruling power of gravity places

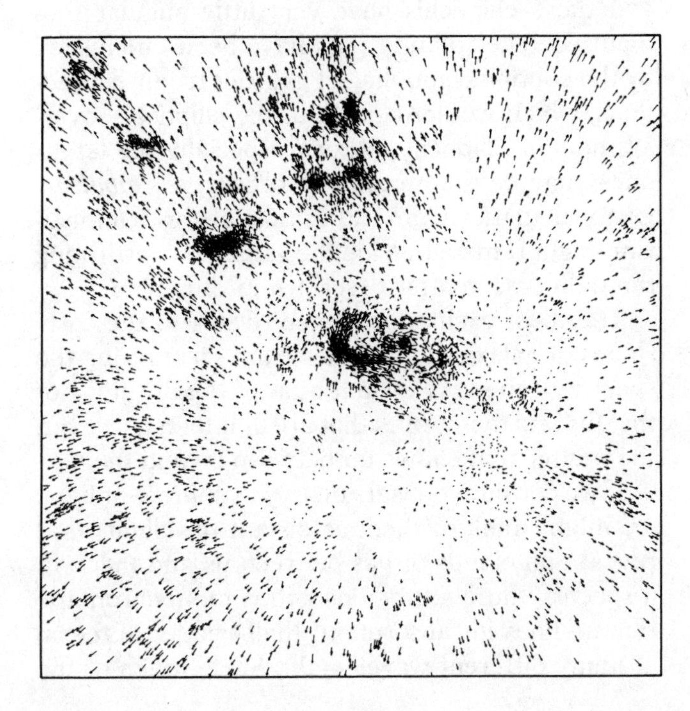

Simulation of clumps in the formation of a protogalaxy and cluster, with dark matter present. (Courtesy Marc Davis.)

protogalaxies in a complex standoff. On one side, there is matter (irrespective of its form) that fails to be pulled into galaxies; on the other lie protogalaxies, some with widely varying amounts of matter. The situation does not remain unchanging since gravity makes the players take ever-changing positions. Some galaxies are gobbled up by others, and eventually one giant galaxy may become a dominant member in the cluster of protogalaxies. The other galaxies move, from the play of gravity, and some settle into an uneasy position against this giant galaxy, balancing its pull with their own motion. The matter that doesn't get pulled into the galaxies tends to stay as a diffuse cloud, permeating the entire cluster of galaxies. (See Plate 18 following page 142.)

The major characteristics of this computer simulated scenario do reflect the galaxies as we actually see them. In many clusters, there are giant elliptical galaxies many times more massive than others (estimates are made with the virial theorem), and these ellipticals have very little angular momentum. The rotation may have been diminished by the continual gobbling of galaxies from all directions, which would cancel out the ellipticals' overall motion. Elliptical galaxies, especially the larger ones, may be the product of *galactic cannibalism.* Leftover matter from the victims of this cannibalism might remain outside the galaxies, constituting the unlit dark matter described in Chapter 3.

The most exciting evidence relates to the early stages of galaxies and comes from scrutinizing the spectra of the very distant quasars. Oddly, it is not the quasars themselves that are of interest here but the matter that shows up between us and them.

If we consider quasars merely as sources of light, we might think of them as distant flashlights that reveal thin clouds of gas between us and them. In a spectra, these gas clouds act as monochromatic filters—lines in "absorption" (half-opaque screens) at many different wavelengths. Since much of the

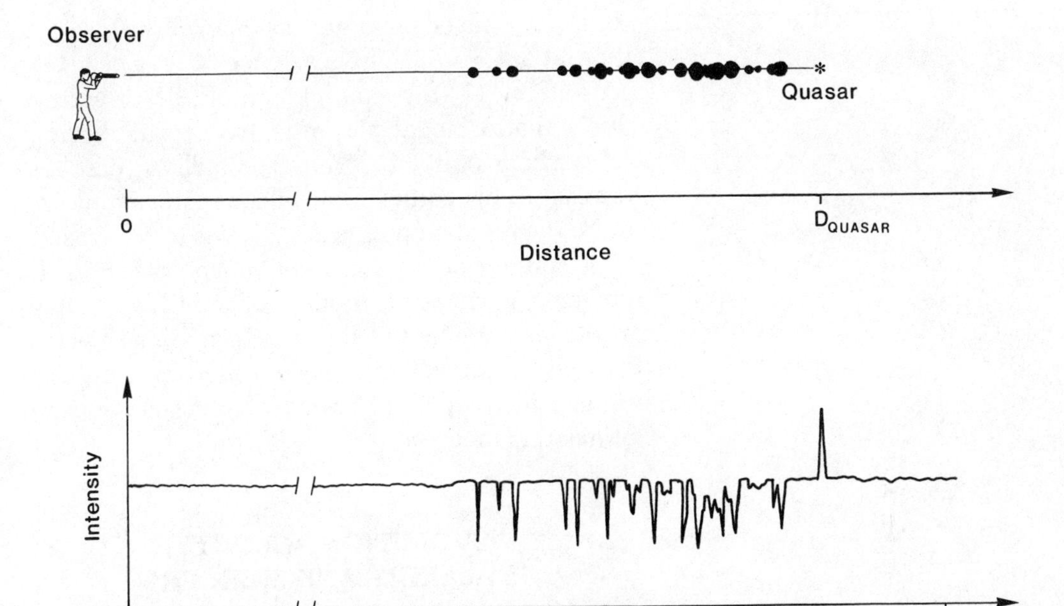

Idealized portrayal of a Lyman α (alpha) forest and a schematic of the relative positions of gas clouds that produce the absorbing spectral lines. Here, the quasar has a redshift of 1.555.

known matter in the universe is hydrogen, these absorption lines are likely to be the result of atomic hydrogen spectral lines, and their different wavelengths imply hundreds, if not thousands, of different tenuous gas clouds, many close by to the quasar. Most of the lines, called the Lyman α (pronounced *alpha*) forest lines after the strange spectral appearance of the Lyman α spectral line, are at redshifts only slightly smaller than the quasars' spectral lines. The easiest way to explain these lines is that they are the tracer of the leftover gas clouds that congealed together to make up the protogalaxies; the absence of a Lyman α forest for nearby galaxies hints at the special environment of the very distant quasars. The quasars may be early galaxies whose environment is still dominated by nearby gas clouds or near-neighbor galaxies.

To astronomers, the process of forming galaxies is not controversial, but the time scale is. Most theorists believe that it takes billions of years for the protogalaxies to form from seeds. This means that we should never see a quasar at a redshift greater than about 5 to 7. The existence of a quasar at a redshift of 12 (for example) would be direct evidence that protogalaxies must have formed earlier on, perhaps only a few hundred million years into the Big Bang, rather than billions. This would require the formation of protogalaxies to be "biased," as if some seed was present much earlier in the universe, perhaps at the time of the CBR. Perhaps a smaller-scale hint of how the formation of galaxies can be controlled may be found by viewing a similar phenomenon of much closer galaxies—that of galactic mergers.

EVOLUTION: GRAVITY, GALAXIES, AND MERGERS

The processes of galaxy formation and change all depend on gravity, as well as on the passage of a great deal of time, perhaps several billion years. As such, the growth stages of a galaxy depend on processes that are universal and slow.

The situation a galaxy finds itself has profound effects on its ultimate appearance. Astronomers speak of a galaxy having an "environment." Does it have neighbors? Will it be destroyed by being drawn into a giant elliptical? Observations provide direct evidence that the environment of a galaxy is the predominant factor in how it grows and how it looks to us. The environment refers to how close other galaxies might be. Not surprisingly, there are many aesthetically intriguing cases where close neighbors make for aberrant appearances.

The way these close-neighbor effects take shape is through distortions of spiral arms, rings and bridges of matter, and violent active-galaxy activity in the core. Galaxies can merge to some strange-looking new shapes. (See Plate 17 following page 142). Trails can run along billions of kilometers, telling the tale of a galaxy liberating its angular momentum from the gravitational attack of a neighbor. The core of a galaxy may have to respond to a neighbor whose distance may be a million times farther away than the actual core's size.

It may not seem to make sense that the core of a galaxy must react to a neighbor, but again it's a matter of angular momentum. If a galaxy (a spiral, for instance) has settled down over a billion years or more, it has "relaxed"—it has allowed all its stars to come on track to a systematic, controlled rotation about the core. Yet when another galaxy slowly approaches, it generates a tide and the core is thrown into a state of turmoil as the orbits of the outer spiral arms get disrupted. Stars in the core collide, then pull together, eventually forming supermassive stars that ultimately collapse and become black holes. With all this disruption going on and gas being drawn into the black hole, the core becomes active, giving the new appearance of an active galaxy.

In 1987, American astronomer Susan Simkin and her colleagues discovered that an active galaxy, MK 348, had a neighboring galaxy that exerted a gravitational stress upon it. With 21-centimeter hydrogen line observations, the active galaxy was shown to be a spiral galaxy whose arms were too dim to show up in a visible light image. The neighbor had caused this spiral galaxy to turn on an active phase in its core. (See Plate 16 following page 142.)

How does all this affect the notion of galactic evolution? The explanation remains somewhat tentative because of the need for more observations. But one simple idea is consistent with what we see, and it goes like this:

1. Protogalaxies form from seeds, which pull together a sea of matter, causing clumps that vacuum in yet more surrounding matter.
2. Every protogalaxy encourages the growth of other protogalaxies, which in turn produces clusters of near neighbors dominated by cannibalistic giants.

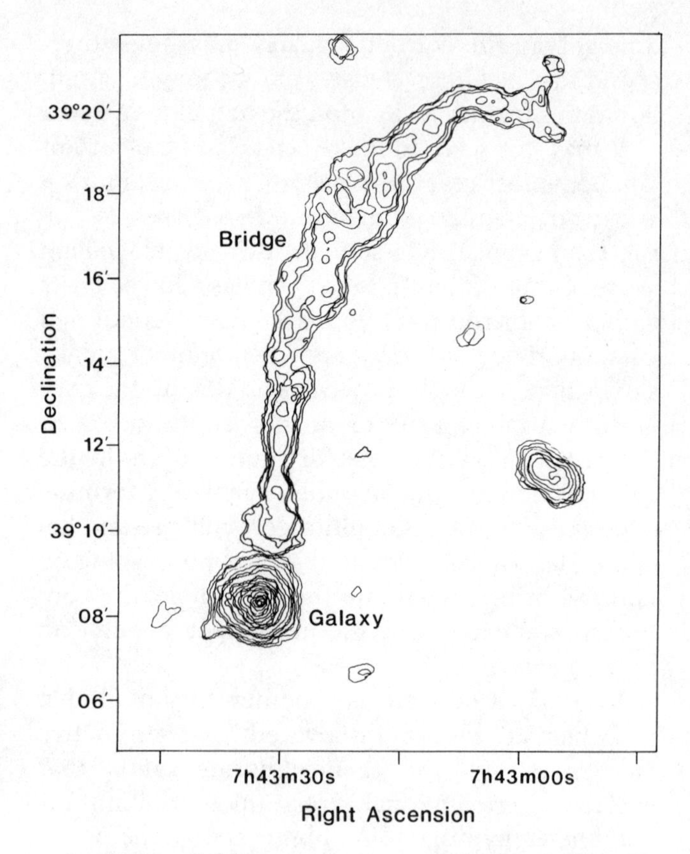

A huge trail of gas emanating from a galaxy distorted by the presence of a nearby companion. (Courtesy P. N. Appleton, F. D. Ghigo, J. Van Gorkom, J. Schombert, and C. Struck-Marcell.)

3. Near neighbors cause protogalaxies to disrupt their own structures, forming cores that draw in hot gas and thereby turn on a quasar phase for protogalaxies.
4. Quasars last a few hundred million years, as they burn up their fuel. They eventually become normal galaxies, mere ghosts of their previous brilliance.
5. Occasional close encounters of neighboring normal galaxies cause mergers and/or a relighting of the quasar phase, giving us active galaxies. (See Plate 15 following page 142.)

Quasars are thus early stages of normal galaxies, the adolescent phase of galactic evolution, when galactic neighborhoods were young and overpopulated. They are so distant that the cores, albeit intrinsically bright, appear as mere pinpoints, whereas the closer but dimmer active galaxies reveal some of their cores' details.

The implications of this scenario to cosmology are universal and at the same time restricting. Unless a galaxy is isolated—with no near neighbors—it may be active, giving a false indication of the brightness expected from a more normal state. Virtually every galaxy used in a cosmological test then needs to be checked individually—through identification of neighbors or through Lyman α forest spectra—to check for the brightness changes wreaked by the cosmic ecology of a crowded neighborhood. And its size and shape may be modified by the presence of neighbors, too. Galaxies, in general, are tracers not just of the expanding universe but of a lifetime influenced by the closeness of crowds. (See Plate 20 following page 142.)

· SIX ·

Micromegas—Inflationary Universe and the Particle Zoo

THE elementary particles—the fundamental and infinitesimal bits of matter in the universe—may seem insignificant to the study of cosmology, which deals in huge scales of size and mass. Yet we must not forget that in the very beginning of the universe, it was these particles that were dominant. Here we shall see that the elementary particles and the forces they convey provided the foundation for the Big Bang, producing and guiding the expansion from the first moment and later shaping its forms.

BIG BANG PROBLEMS

Despite all the knowledge that originally led us to accept the validity of the Big Bang model, there remained, until recently, certain key aspects of the universe that didn't jibe with the scheme.

First among these problems was the form of "stuff" in the universe. In 1928, Paul Dirac, a French physicist, showed that the subatomic particles have counterparts—antielectrons, or positrons, for example. The positron is the same as the electron and reacts the same way, but its charge is the opposite; while the electron is negatively charged, the positron is positively charged. Dirac showed that the positron was but one of a family of antimatter particles that must exist and that the joining of matter and antimatter leads to a spectacular energy burst through mutual annihilation. Discovery of antimatter happened in the laboratory, shortly after Dirac's prediction, and today it is well known that high-energy cosmic

(a)

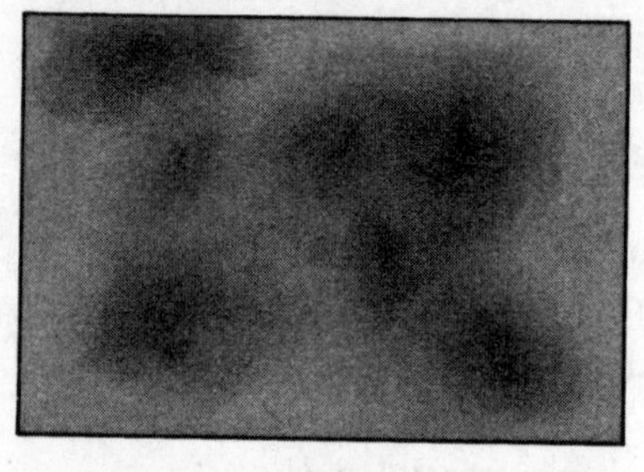

(b)

The horizon problem—the early universe looks very smooth (a) instead of full of gross fluctuations (b).

rays from deep space can break down into pairs of electrons and positrons spontaneously.

This discovery was carried to its logical conjecture—that there exists an antiuniverse, a universe composed entirely of antimatter. In fact, we find antimatter only rarely; this is lucky for us, for equal and large amounts of matter and antimatter give rise to a universe of brilliant destruction. Astronomers refer to this imbalance—how much of the universe is matter versus antimatter—as an *asymmetry.* For there to be this kind of asymmetry, the Big Bang must have favored the production of regular matter—but why did it do that?

A second problem astronomers faced with the Big Bang was explaining why the CBR is so smooth. This mystery goes back to one of the first assumptions in the Big Bang theory, that of isotropy. Gamow surmised that the CBR would be isotropic, but no one was able to explain why the isotropy must happen. The relative smoothness of the CBR (which, as we shall see in Chapter 8, is not absolute) suggests that the earliest universal stage must have formed so quickly that parts of it could not have had time to deviate from the others in temperature. The dilemma of justifying the CBR's smoothness is often called the *horizon problem.*

Yet another enigma has been labeled the *flatness problem,* this one arising from the theory that at ρ_{CRIT}, the critical density, our universe would be closed. Current observations would indicate that the universe today weighs in at only 15% of the critical density. Under a simulated reversal of time, however, it is possible to show that the 85% difference becomes but a fraction of this at the earliest moments of the Big Bang explosion, when all parts were much closer together; at that time the universe's total volume was far smaller. At its beginning, then, the density of the universe must have been at or near the critical density point, even if it does not appear so today.

The answers to these problems are not apparent

in the standard Big Bang model, which reveals the nature of the universe after its infancy, when the force of gravity took control of the universe. But gravity is not the only force in the universe; the Big Bang has a beginning where the other forces are given a role at least equally as important and where the elementary particles dominate the workings of the universe.

FORCES AND THE MOMENT OF BEING

Throughout the decades, laboratory physicists have wondered how matter holds together and have described this in terms of different forces through the various subtleties of attraction and repulsion. Gravity is one of these forces, while electromagnetism (radiant energy, such as light) is another. These forces can act over great distances, but there are at least two other forces of nature that are less important than the others except at very close distances.

These two forces are called the *strong* and *weak* forces. Both exert their effects over minute distances less than 10^{-15} meters (much less than the size of an atom).

The strong and weak forces are not apparent to us in our everyday experience. In fact, outside the contrived environment of some laboratories, they

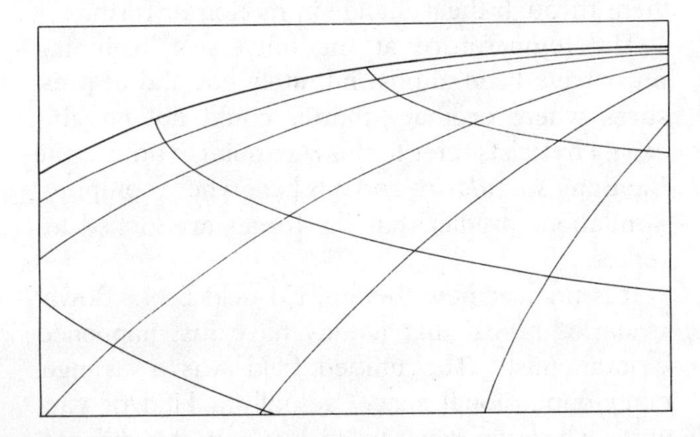

The flatness problem—the universe expanded so quickly that its curvature was flattened, much as the curved Earth looks flat to us.

Table 6.1 The forces of nature and the distances over which they are important

Force	Distances
Electromagnetism	All
Gravity	All
Strong	$\sim 10^{-15}$ meters
Weak	$\leq 10^{-17}$ meters

are incredibly difficult to detect. In the infant Big Bang, however, the universe was packed so closely together that these two forces were significant. To a physicist, even a simulation of this environment provides an opportunity to see how all the forces work in concert.

Albert Einstein was among the first to ponder how the forces of nature are tied together. He spent over 20 years, until his death in 1955, trying to unite electromagnetism with gravity in his unified field theory. Although never successful, he set the stage for later work that was able to take into account the strong and weak forces.

Within the last generation, physicists have used their knowledge to pursue a *grand unified field theory* (GUTS). GUTS incorporates a family of models for bringing together all the forces of the universe by relating how force *fields* form and how elementary particles convey the forces as they transmit them through their change in motion or form.

The temperature at the universe's beginning must have been almost infinitely hot and at pressures where ordinary matter could not possibly exist. Physicists refer to this zero point in time as the Big Bang *singularity,* and it is here where computer simulations predict that the forces are locked together.

It is unclear how this unified field broke down; scenarios reveal that it may have just happened spontaneously. The unified field was a strange, many-dimensional space, actually a kind of vacuum, where energy existed but particles did not.

As the unification broke, the vacuum became a flurry of new, elementary particles, called *quarks,* that were undergoing a change of state. According to this model, as the unity broke down, the new universe went through a *phase transition,* not unlike that of an ice cube melting to water. But this phase transition had awesome consequences. First, a plethora of other elementary particles was created from the combining of quarks, and the preference of one force over another led to the making of much heavier particles—baryons—such as protons. In an instant of time, 10^{-35} seconds, antimatter was created, but only in small amounts, because the temperature was more conducive to forming baryons—regular matter. Most of the quarks locked in to making baryons; quarks continued to exist only as parts of heavier elementary particles, only a few made up antimatter.

INFLATION—THE RUNAWAY UNIVERSE

By far the most enlightening aspect of this unified force model is the great amount of energy released, energy that would explain why the universe exploded in its Big Bang.

In 1979, American physicist Alan Guth was the first to realize that the energy released during the breakdown of unification would cause the universe to expand radically. He showed, based on an analogy to other phase transitions, that this breakdown would occur over a remarkably short space of time, perhaps 10^{-32} seconds! So much energy released over a short time would trigger an expansion that would double the size of the universe every 10^{-35} seconds. He dubbed this rapid expansion *inflation,* and today astronomers refer to the earliest moments of the Big Bang as the *inflationary universe.*

Alan Guth, MIT physicist who was the principal founder of the inflationary model. (Courtesy MIT news office.)

This infant universe was expanding so rapidly that most of its expansion happened over a fraction of a second. Only when the phase change was completed did the energy output trail off, and the expansion settled down to something closer to what we are familiar with. It is important not to lose track of the relevant sizes, too; starting as a mere point, the universe was only the size of a soccer ball when the inflationary stage lessened. After that, the expansion continued as a follow-through of this previous, if only momentary, extreme, and new types of particles formed as the forces became separate from one another.

The inflationary aspect of this phase transition provides a solution to the antimatter problem, as well as to the horizon and flatness problems. The horizon problem, the smoothness of the universe, can make sense if the universe was uniform when the symmetry breaking happened. Inflation would have caused the parts to expand so rapidly that they would be a blown-up, albeit less dense, replica of the original state. Inflation would freeze in the characteristics of the time of symmetry breaking, when everything was uniform. Inflation is the reason that the universe has such a near-perfect isotropy.

Ice undergoes a phase transition; it stays at a constant temperature until it has changed its phase from ice to water.

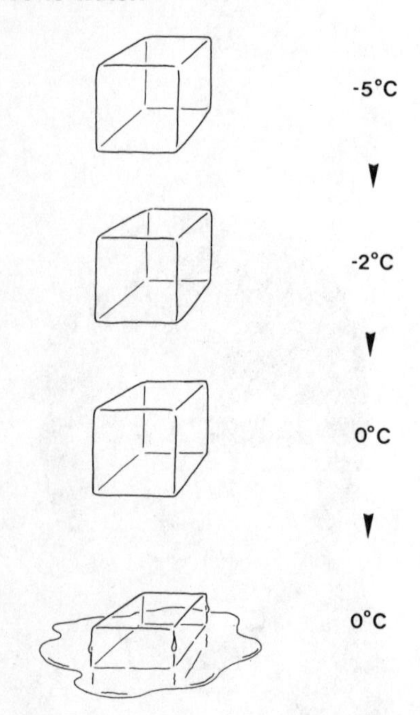

The flatness problem can also be explained by inflation. Inflation happened so fast that all curves in space were quickly made less severe; they were virtually flattened. Another way to look at this is that the rapid expansion caused enough matter to be made so that curvature was wiped out. Since curvature is zero, or flat, when the universe is at the critical density, this implies that inflation must occur for an uncurved universe: $q_0 = 0$. In a very real sense, the process of inflation demands the universe to be closed.

COSMIC STRINGS—DEFECTS FROM INFLATION

When ice forms from water, slight impurities cause the production of defects in the crystallized ice. In a similar way, the slight randomness in the early, unified universe causes imperfections that show up

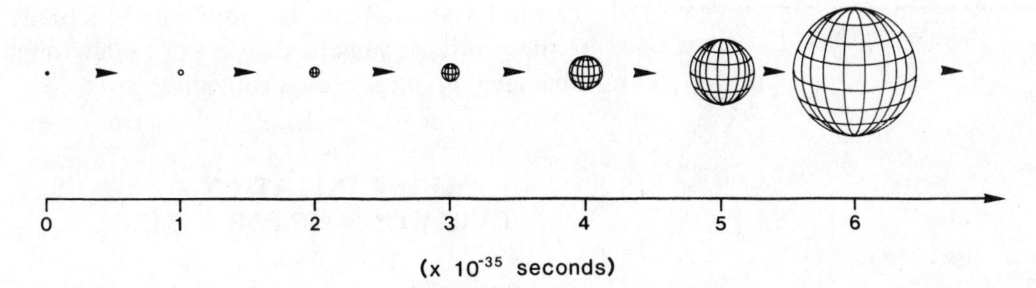

As the universe underwent its phase transition when the unified forces decoupled, its size increased, doubling about every 10^{-35} seconds.

as defects later on. What evidence do we have of a defect in the early breaking down of symmetry of the forces? The answer lies in *cosmic strings,* first postulated by the Russian-American physicist Alexander Vilenkin in 1977. These are bits of isolated space where there were defects in the symmetry breaking. Before they ultimately evaporate, cosmic strings appear as cracks in space and time.

Perhaps cosmic strings might be better described as pure bits of unified force fields. But to us, they would look like odd masses of matter. Incredibly narrow, only a few atoms wide, they may form long loops billions of kilometers long. Long yet slender, they have strong gravitational fields that can greatly attract the mass around them and distort the light in back of them.

These strings may form at symmetry breaking, but only a few could last for a long period of time. But for those that do last, theories suggest some fascinating properties. American astronomers Jeremiah Ostriker and Ed Witten have found an even stranger twist to cosmic strings: They may not just attract matter; they may also repel it. This is caused by another property of strings, their strong radiating tendency. This radiation can be intense enough to push away all matter near the string, effectively clearing out a channel of space for billions of miles. Another possibility is that strings jitter about the universe, zipping along at the speed of light. Could a cosmic string ever encounter Earth, wreaking havoc from its strong gravitational pull? The only thing we know for certain about cosmic strings is that they make sense in the computer simulations, but until one is actually seen, these other conjectures will not be anything but challenging intellectual problems.

AFTER INFLATION—EVOLUTION OF PARTICLES

Only a fraction of a second after the beginning of the universe, the inflation stage ended. But the next few minutes' slower expansion established what the universe would be like for the billions of years to come.

During these moments, the expansion caused the temperature and pressure of the universe to drop dramatically. As the computer simulations reveal, the quarks of the symmetry period had already been transformed into a sea of electrons and positrons—matter and antimatter—which would quickly die and annihilate each other. After one-hundredth of a second, the universe became a bath of light and small particles called *neutrinos.* By one second after the expansion, the temperature had fallen to 10^9 degrees, only a few times hotter than the temperature in the interior of stars. These conditions favor the manufacture of protons and neutrons, with many more protons than neutrons.

As the expansion continued, the temperature of the universe dropped, producing certain types of particles for very short times. By three minutes into the Big Bang, the neutrons and protons had combined into the simplest form of hydrogen—an atom of a neutron and a proton called *deuterium*—but most of the protons and neutrons combined to form helium. Atom manufacturing came to a halt as the rapidly cooling temperature prevented more combinations. By an hour into the Big Bang, almost all production of new particles had ceased.

What remained was mostly a universe of energy and matter, in the form of photons, electrons, protons, deuterium, and some helium. Neutrinos, formed at the very beginning, continued to exist in copious numbers.

As the universe continued to expand, its temperature dropped precipitously. Eventually, as the pressure fell, the energy-dominated universe became one from which light could escape, to be followed by particles dominating the scene. What had started as a unification of fields became a sea of energy and elementary particles and, eventually, a universe of hot baryons that formed simple atoms such as hydrogen and helium.

NUCLEOSYNTHESIS—THEORY MEETING OBSERVATION

This "cooking" property of the universe—its ability to make certain particles at different times in its early history—was one of the main reasons that Gamow was motivated to construct the Big Bang model; he believed it explained how the elements were made. Yet Gamow extended the scenario a little too far; the elements much heavier than helium were not made in the Big Bang. In fact, the kinds of temperatures and pressures needed to make heavier elements, such as carbon and oxygen, were not to be found for more than a fleeting moment in the Big Bang. In 1957, American astronomers Geoffrey and Margaret Burbidge and William Fowler, along with British astronomer Fred Hoyle, showed that supernovae and the insides of stars were more fertile grounds for cooking together the heavier elements. Today, it is believed that the simplest elements—hydrogen, helium, and lithium—were mostly made in the Big Bang, while the *nucleosynthesis* of heavier elements—everything heavier than lithium—happened in stars, or in supernova explosions.

It turns out that the amount of deuterium that was made, compared with that of helium, depends on the specific densities that existed during the

Plate 15 A distorted galaxy, torn by the merger with a neighbor. (*Courtesy NOAO.*)

Plate 16 MK 348, an active galaxy whose visible core (pink) is dwarfed by spiral arms that show up as 21-centimeter hydrogen (blue). This spiral galaxy has a neighbor that is making it active. (*Courtesy NRAO.*)

Plate 17 The toadstool—image-enhanced photograph of merging galaxies. (*Courtesy NOAO.*)

Plate 18 A cluster of galaxies in the constellation Hercules, with foreground red and blue stars in the Milky Way. (*Courtesy R. Schild, SAO.*)

Plate 19 The product of a mortal explosion, the supernova SNR 1987a (upper center) shines brightly. This is the closest supernova (lying just outside the Milky Way) to occur in centuries, giving us an unprecedented chance to study this cataclysm in a most comprehensive way. (*Courtesy NOAO.*)

Plate 20 A galactic cluster in the constellation Coma Berenices. (*Courtesy NOAO.*)

Plate 21 Computer simulation of the Swiss cheese lumping of galaxy clusters, produced through the presence of dark matter—massive neutrinos; cosmic strings may not be the only way to get a Swiss cheese universe. (© *Joan Centrella.*)

Plate 22 Out to a distance of 340 megaparsecs, galaxies supercluster in Swiss cheese-like structures, as shown by this study by Brent Tully. Bubbles and voids are just another way of describing this type of structure. (*Courtesy Brent Tully.*)

Captions continue on page 143.

Plate 15

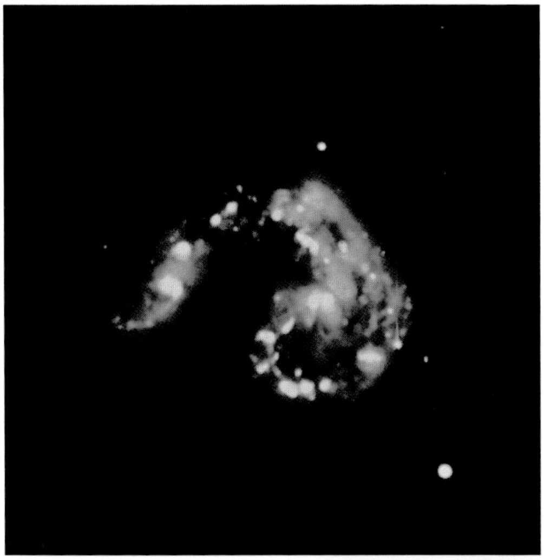

Plate 16

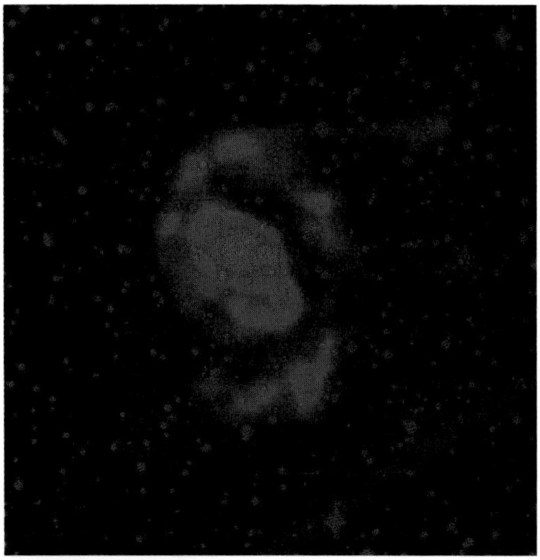

Plate 17

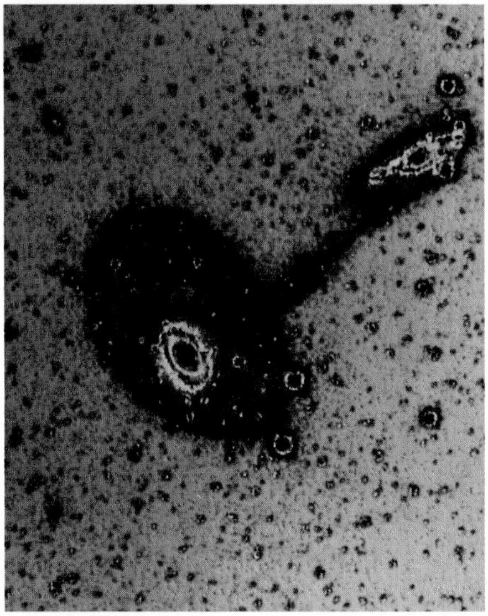

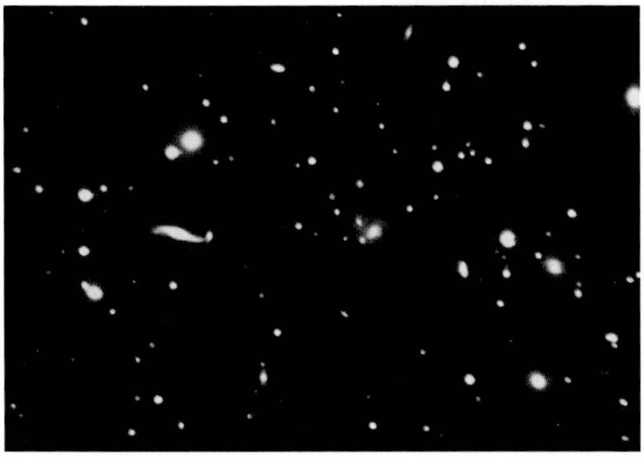

Plate 18

Plate 19

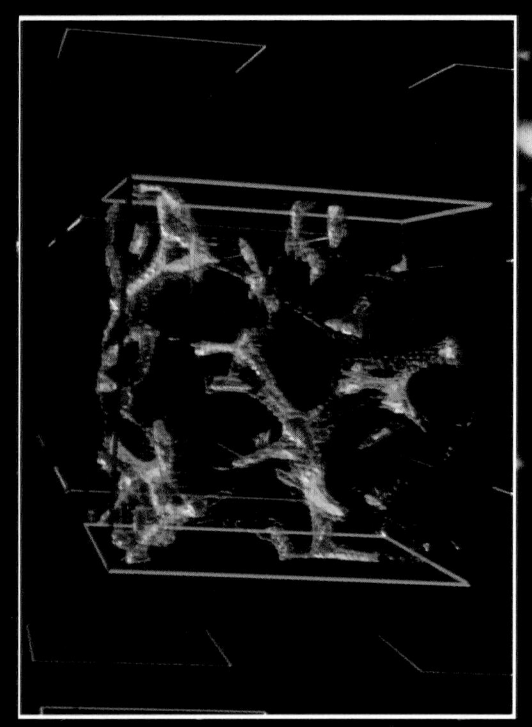

Plate 21

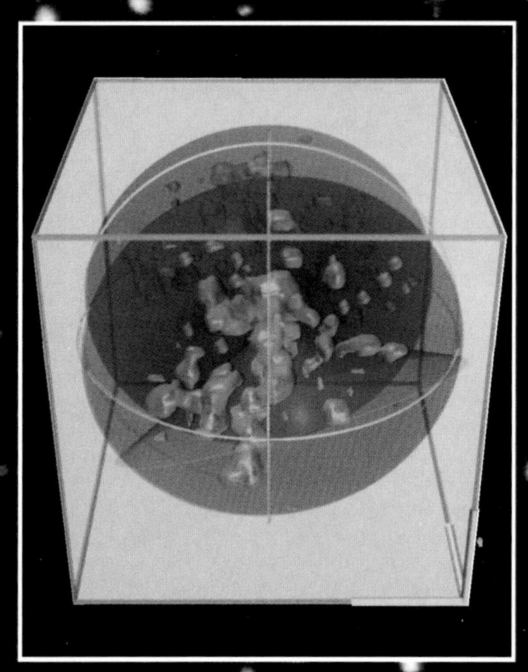

Plate 22

Plate 20

Plate 23

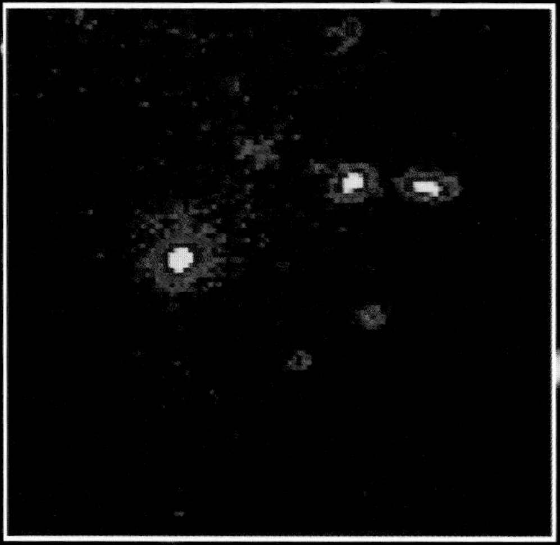

Plate 24

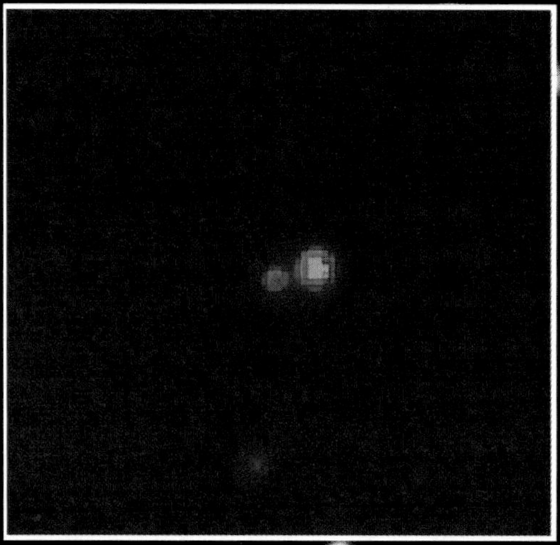

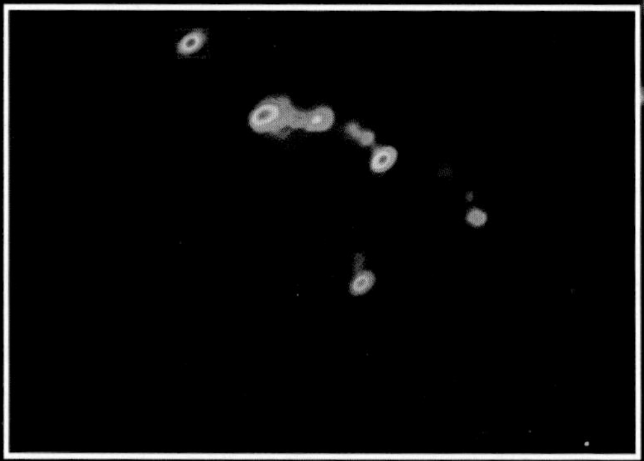

Plate 25

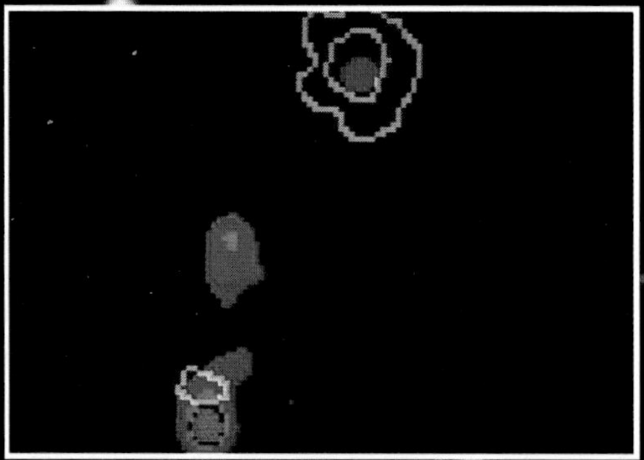

Plate 26

Plate 27a

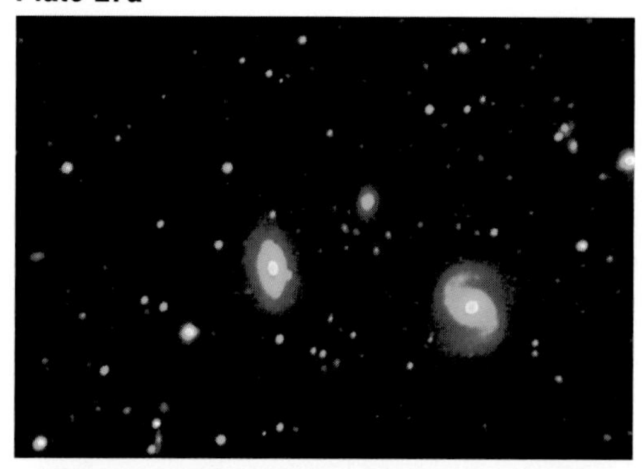

Plate 27b

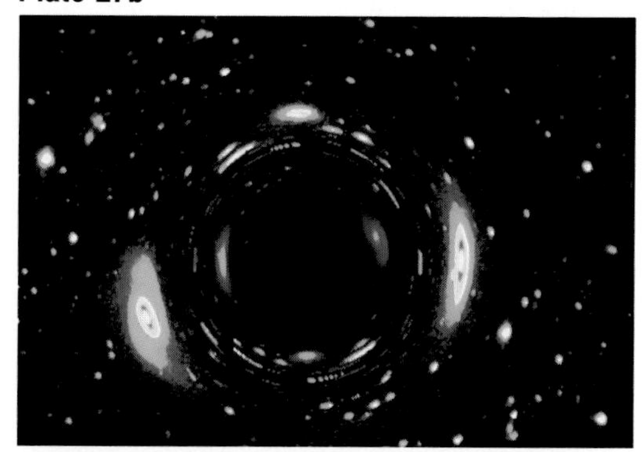

Plate 28

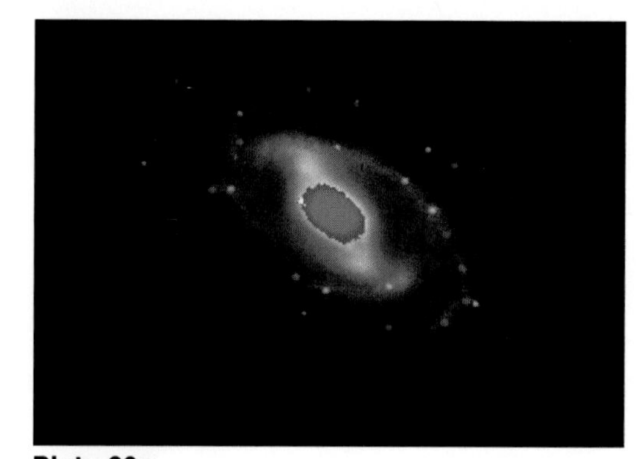

Plate 29a

Plate 29b

Plate 23 False-color CCD image of a quasar (center, left) looks fuzzy compared with the image of a star (upper right). The quasar is not point-like, but extended like a galaxy. (*Courtesy M. Malkan.*)

Plate 24 UM673—two visible-light false-color images (center) separated by about two arcseconds in angle. The bottom object is a galaxy, unrelated to the bender, which has been electronically removed from this photograph to show the two images. The bender is a galaxy lying between the two images. (*Courtesy Jean Surdej and ESO.*)

Plate 25 False-color radio photograph of the Twin Quasar, showing the radio structure of A and B. (*Courtesy NRAO.*)

Plate 26 Gravitational lens 0023+171. Here, the radio image (blue) coincides with the optical ones. The center (blue) radio source may be the bender, but its visible-light counterpart is absent. (*Courtesy NOAO.*)

Plate 27 A computer simulation of a galaxy cluster being gravitationally lensed. (a) cluster without a black hole (dark matter) bender and (b) with the bender. (*Courtesy E. Falco.*)

Plate 28 Cosmic arc Abell 370. This gravity megalens must be formed by a great amount of dark matter in the bender cluster. (*Courtesy NOAO.*)

Plate 29 In this simulation, a galaxy (a) has been placed in back of an invisible bender; the result is a multi-image mirage of the galaxy (b). (*Courtesy E. Falco.*)

time of deuterium production. These densities are nothing mysterious; they are the density of the universe, already discussed in the Big Bang model. If the universe is closed, the amount of deuterium to helium will be rather small, as more protons and neutrons will combine to make helium. But in an open universe, with a density far below the critical density, the amount of deuterium will be very large.

Deuterium can be found in stars and gas clouds, leftover product of the Big Bang. It emits ultraviolet spectral lines whose intensity can be compared with those of helium to figure out the relative amounts of helium and deuterium. These spectral-line observations reveal that there is a large amount of deuterium around the universe. But its amount reveals that the critical density of the universe must be higher than the observed density, although the uncertainties may be 30% or more.

This finding—that the deuterium abundance implies that the universe is open—is a startling contradiction to the inflationary model, which requires that the universe be closed. However, there is an assumption here; most of the matter ended up as baryons rather than other types of more exotic particles. Perhaps this was not the case, and the neutrinos or other particles might have constituted the bulk of the matter made in the Big Bang. In this case, the (possibly) massless neutrinos would not make up the universe of galaxies and light but remain nearly invisible. Alternatively, the cosmic strings could have been made in such high abundance that they are a factor in closing the universe. Clearly the Big Bang could have thrown its bulk into particles that do not make up the visible universe but produce strange forms of dark matter never to light up the sky as galaxies of stars, invisible remnants of a noble beginning.

· SEVEN ·

Gravity's Lens— The Cosmic Mirage

THE universe is controlled by gravity, the force that binds its parts together. The universality of gravity allowed Einstein to derive his field equations and LeMaitre, Gamow, and others to establish foundations for understanding the origin and evolution of the cosmos. But underlying Einstein's field equations was the notion that gravity's curvature of space and time created a bizarre effect—light rays could be deflected and bent. Gravity didn't limit its attraction to matter. It controlled the paths of light rays, too.

Long before Einstein, in fact, Isaac Newton had predicted this phenomenon. In 1687, Newton proposed that light had weight and that light rays behaved as a collection of minute particles. As a light ray skimmed past a massive body, Newton believed, the light ray would be attracted and its path bent. Yet even Newton surmised that the amount of the deflection was too small to measure with the equipment of his day.

More than 200 years later, Einstein reconsidered the bending problem. Einstein's framework was vastly different. To him, light rays traveled along paths, called *geodesics,* in the dimensions of space and time. But light itself does not have weight. As space-time became warped by the gravity field of a massive body, the geodesic bent, too.

Einstein's first estimate of the degree, or angle, of deflection was flawed, and the angle he derived was identical to that predicted by Newton. Here was a bit of a problem—a theory that predicts the same result as another, better-established one lacks a compelling reason for its acceptance. Yet, upon additional work on his theory of gravity, Einstein found the bending

Isaac Newton (above), who came up with the classical theory of gravity, in which light rays bend because they have weight. (Courtesy AIP, Niels Bohr Library.) Albert Einstein (right), whose general relativity maintains that light paths are bent from the distortion of space by a mass. (Courtesy Caltech Archives.)

with his model was actually twice as large as Newton's prediction. And with the advent of modern telescopes, Einstein realized that it would be possible to distinguish between Newton's model for gravity (light has weight) and his (light is weightless; space-time gets distorted or bent) by means of observations.

In 1913, Einstein asked the prominent Mount Wilson astronomer George Hale for his advice on observing the deflection. Einstein pointed out that the sun's mass might bend light in such a way that particular stars might have slightly different relative positions at times when the sun was near the stars and at times when it wasn't. Hale pointed out that a solar eclipse—when the moon blots out the sun temporarily—might be a superb opportunity to search for a deflection of stars lying close in angle to the sun. In a later calculation, Einstein showed that the deflection could be as large as 1.8 arcseconds for stars very close in position to the sun during the eclipse. Newton's prediction would be half that value. This angular change could, in theory, be easily measured—but only under the best of sky conditions.

In 1919 the British physicist Arthur Eddington led an eclipse expedition to investigate this possibility of light bending. To find the perfect conditions, he and his colleagues made a demanding journey to remote, bleak places in the South Atlantic: Principe in West Africa, and Sobral in eastern Brazil, two extreme points lying along the path of the solar eclipse. The research was hampered by intense heat and humidity, as well as the ubiquity of native insects, which were not beyond tasting photographic emulsion (and researchers' flesh) for its potential palatability. Still, a few plates of the eclipse were salvaged, revealing star deflection consistent with, but not quite precise enough to verify, Einstein's predicted deflection. More plates were required for proof.

Three years later, an eclipse expedition by astronomers at the Lick Observatory in California made Einstein's general relativity theory for gravity the new, and more accurate, paradigm. In a remote desert in Australia, the moon's shadow made a spectacular eclipse, showing over 140 bright stars near the sun's position. So many stars were deflected, and measured, that the conclusion was overwhelmingly in favor of Einstein rather than Newton. It is this result that catapulted Einstein to his reputation as a rare genius.

THE GRAVITATIONAL LENS

Einstein was aware that the bending of light rays by the gravity of a massive object is similar to the refraction of light by an optical lens, such as a magnifying glass. But it wasn't until the 1930s that he

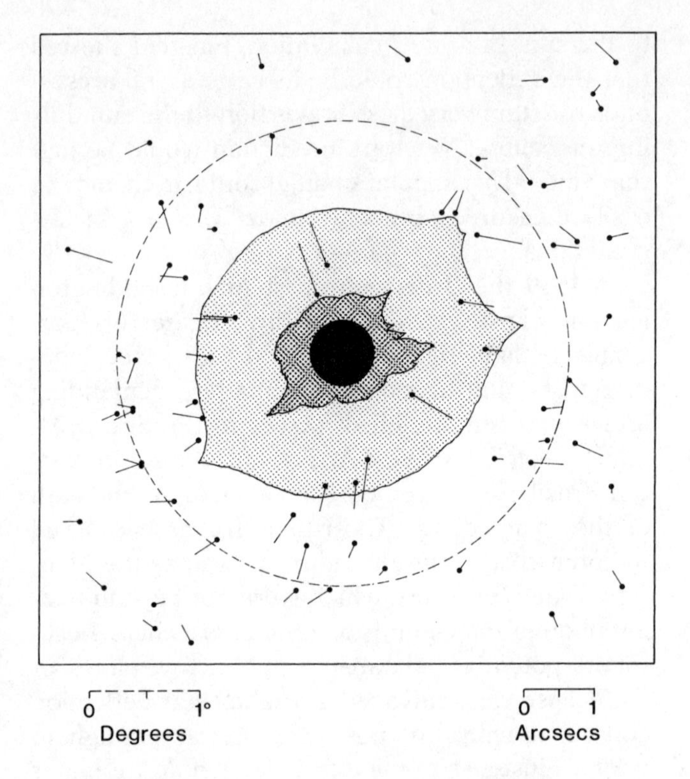

The deflection of stars during the 1922 solar eclipse. The sun (center) is surrounded by its corona; the outer circle represents an angular scale from its center. Lines represent magnified indicators of the amount and direction of deflection of individual stars. (From the Publications of the A.S.P. Vol. 35, p. 158. © 1923 Astronomical Society of the Pacific.)

extended the deflection problem to the lens analogy to study how gravity can form and magnify images of the bent light.

The lenslike action became apparent to Einstein as a thought problem. If two distant stars, each mere points of light, are aligned with an observer, what happens to the background star—does it disappear, or does something else occur? Einstein showed that both stars should be visible, but the background star would not appear as a point of light. Rather it would appear as a very bright ring—now called an *Einstein ring*—with the foreground star at its center. Because the background star's image has been reshaped and magnified, the gravity of the closer star has acted as a kind of lens. This is called the *gravitational lens effect,* or the *gravity lens* for short.

The Einstein ring shows up only when the observer and both stars form a perfect alignment. The gravitational lens effect rapidly diminishes when the alignment is broken. Yet there is a curious result when the alignment is almost perfect. The Einstein ring breaks up into two opposing crescents; two "moons" form with horns pointed at each other. In fact, these are two magnified and distorted images of the background star. As the alignment becomes even less perfect, one of the images disappears and the second becomes dimmer and smaller. Eventually, the alignment is imperfect enough to just deflect the image of the background star, rather than forming a ring or two crescent images.

The bending of stellar images during a solar eclipse is just a special case of a gravity lens. Here, the foreground star—the sun—is not a point on the sky but has a relatively large angular size. Any stars that align with the sun and an observer will not be formed into two crescents or an Einstein ring, because the angular size of the sun is many times larger; the gravity lens effect can be achieved with the sun only by increasing the distance of the observer to several hundred times the distance between Earth and the sun! Yet stars that are not quite aligned with the observer and the sun will be bent, just as was observed during the solar eclipses.

The chance of seeing perfect alignment with two distant stars is small because each star takes up such a tiny angular size (less than one-thousandth of an arcsecond). Because this angular size is so minuscule, the chance of finding a background star in alignment is much less than one in a million. Einstein concluded that the full-fledged gravity lens (with stars) was likely to remain an interesting idea uncorroborated by observation.

In 1937 the Swiss-American astronomer Fritz Zwicky raised an interesting question: Could there be a gravitational lens effect with galaxies? There are far more galaxies in the sky than there are stars in our own galaxy, so the chance of finding an alignment of two galaxies was much bet-

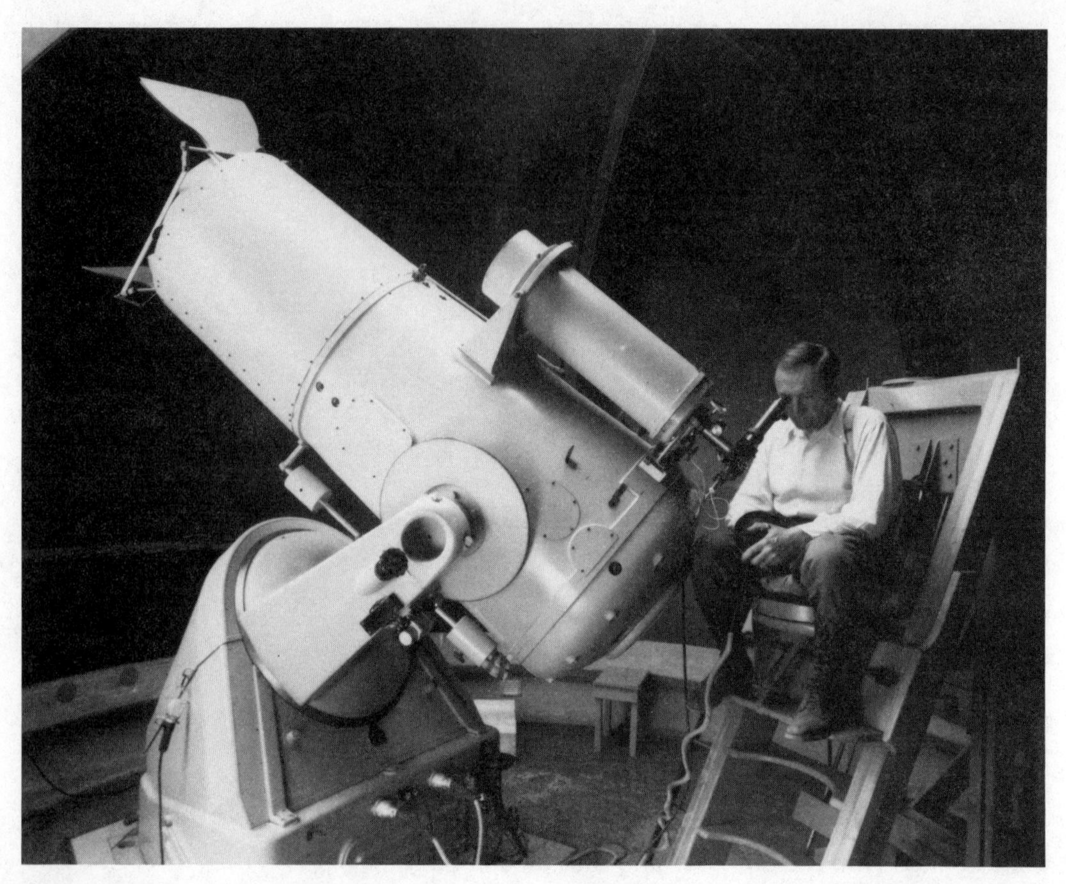

Fritz Zwicky, astronomer who foretold the gravity lensing of galaxies more than forty years before the effect was observed. (Courtesy Caltech Archives.)

ter than a million to one. Indeed, Zwicky asserted, perhaps a little hubristically, that one was bound to find two galaxies aligned and the gravity lens prominent. Unfortunately, Zwicky, a maverick astronomer never quite welcome in the company of such fellow Caltech-Mount Wilson colleagues as Hubble, never had access to the big telescopes he needed to look for gravitational lensing with galaxies. Most astronomers discounted the gravitational lens effect as a fantasy, and the gravity lens remained only an obscure idea, sometimes relegated to the status of an astronomical homework problem. For over thirty years, the gravity lens gathered only humble interest from a few astronomers and physicists. (See Plate 29 following page 142.)

In the mid-1960s, the discovery of quasars

briefly rejuvenated interest in gravity lenses. A biophysicist, Jeno Barnothy, suggested that the peculiar extra brightness of quasars (only a couple of dozen were known at that time) could result from magnification by a gravity lens effect. In his conjecture, each quasar should indicate the presence of two aligned galaxies, one magnified many times by the gravity of the closer one. A quasar was therefore a galaxy made to appear extra bright through a gravity lens. Eventually, too many quasars were discovered—today we know of at least 4,000—for the lens explanation to account for the quasars. Every quasar cannot be a gravitationally lensed galaxy. This doesn't mean, however, that some quasars couldn't be especially bright by a chance alignment with a galaxy.

TWIN QUASAR—DISCOVERY OF A GRAVITATIONAL LENS

Shortly after Barnothy's suggestion, interest in gravity lenses faded again. But then in 1979, a strange pair of objects was spotted by the British astronomer Dennis Walsh and his colleagues. The objects showed up on photographic plates at the position of a radio source with no known visible counterpart. The two objects were only five arcseconds away from each other, and their highly redshifted spectra indicated that they were quasars. Oddly, these two spectra looked virtually identical. Was this Twin Quasar two identical objects or two images of a gravity lens illusion?

When observed at other wavelengths, the odd duplicity of the Twin Quasar held true. For example, the ratio of magnitudes of these twins taken at infrared wavelengths was very similar to the ratio at ultraviolet and visible-light wavelengths. Yet energy is generated quite differently in quasars at these different wavelengths; the chance of seeing two separate quasars with this property of constant

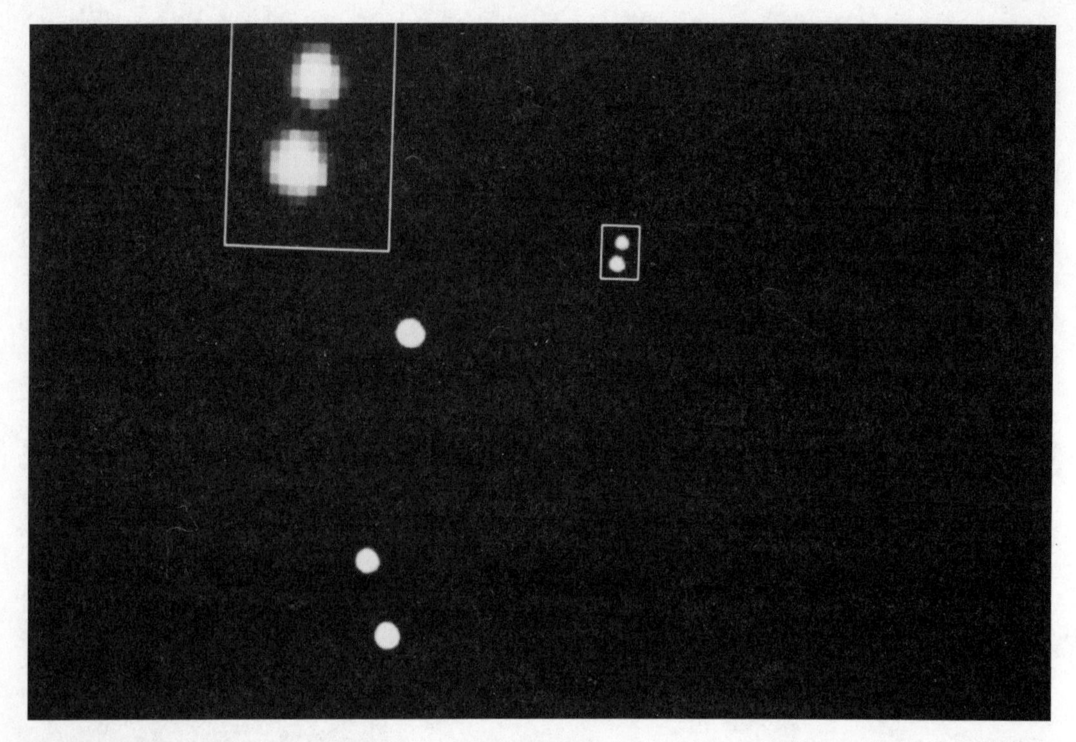

In the constellation Ursa Major, the Twin Quasar shows up as two inconspicuous, closely spaced objects (box enlargement). Other objects here are stars. (Courtesy R. Schild, Smithsonian Astrophysical Observatory.)

magnitude ratios was very low, perhaps a million to one or less.

In a gravity lens, the light bending and magnification depend only on the alignment and the mass of the lensing galaxy, commonly called the *bender.* Also, a gravity lens makes no distinction between different wavelengths when it bends light rays, so a gravity lens gives a natural explanation for the sameness in magnitude ratios at the different wavelengths. The gravity lens seemed the easiest way to explain the Twin Quasar.

This explanation—that the Twin Quasar was the first discovery of a gravity lens in action—was not without early problems. In particular, it was impossible, at first, to find the bender. Within a few months of the Twin Quasar's discovery, however, very sensitive CCD images revealed a host of very dim galaxies near the center of the Twin Quasar. A dim but giant elliptical galaxy lay slightly off center from a line between the two images, labeled *A* and *B,* respectively. The galaxy was the heart of a large cluster of over a hundred galaxies about 2,000 megaparsecs distant. The redshift of A and B indicated that the quasar itself (being "imaged" by the benders) was about 3.5 times more distant.

Complications of Einstein's simple model soon became apparent with the Twin Quasar. These complications were the result of the bender being not a star but a rather large galaxy with other galaxies near it. So instead of having one ring or two crescents, the images of the Twin Quasar were slightly asymmetric and the alignment with the bender (or benders) was slightly off center. This first example of the gravitational lens effect showed how nature always challenges us to understand the complicated special cases, always more complex but often more interesting than the predictions of pen and paper.

WHY GRAVITY LENSES?

In Chapters 2 and 3, I explained that many of the key questions in cosmology come down to finding the values of the cosmological quantities and the solution to a related problem—that of the dark matter. Surprisingly, the gravitational lens effect offers some insight into these problems.

The beauty of this effect is that it depends on three factors. First, the angle of bending depends on how massive the bender is and how its matter is structured—is the bender massive but compact, or is it massive and spread out? Next, the alignment of us as observers with the bender and the background object leads to just how the images look—is the image a ring, or is it two or more different images? Finally, the object will have an angular size—is this angular size very tiny, or is it larger than any ring that might result from the action of the bender? (When the angular size of an object is large, the bender doesn't produce an image at all.)

Equally as critical is an apparently unimportant

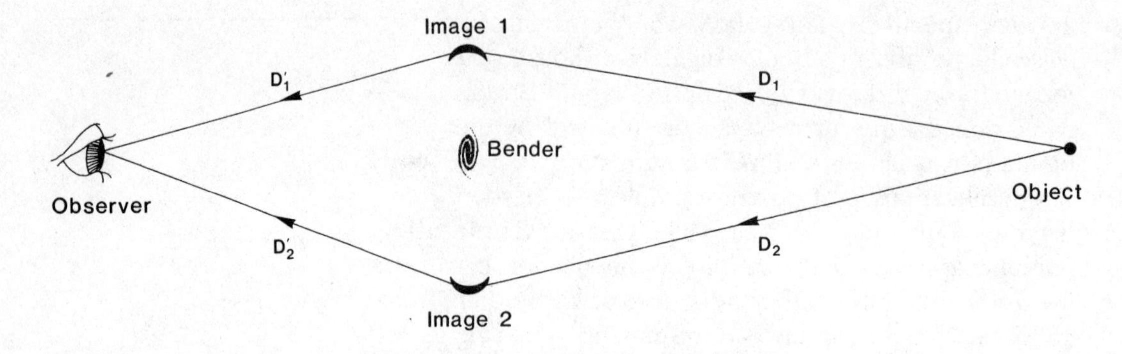

$D'_1 + D_1 < D'_2 + D_2$

Light from an object, such as a quasar, will form different images whose light takes different paths. The paths will be of different lengths, so that one image will reach us before the others; any changes in the object will show up with instant replays at a certain time delay.

factor: the light of the bender itself. Because the bending of the light of the distant object depends on the gravity of the bender, it doesn't matter if the bender's mass gives off any light at all. So the odd mirage of a gravitational lens is a great way of estimating the bender's mass without trying to guess its mass by how much light it gives off. An invisible bender produces the same lens effect as does a brilliant one. Here lies an opportunity to find the mass of a galaxy without worrying about whether you can see it. So the dark matter problem, and other issues of mass in the universe, might be addressed in a totally different way with gravitational lenses.

Another interesting factor of lenses is that the differences between images, in the absence of perfect alignment, give an indication of the distances of the bender and object. Distances, always the key problem in cosmology, cry out for new yardsticks. Lenses are a unique aid as a yardstick because they do not suffer from the problems (such as galactic evolution) that others have. And once distances are known, it is possible to compare them with those that Hubble's law infers, via the redshifts. With distance and redshift in hand, it is possible to calculate H_0 and q_0.

The method of estimating distances from a gravity lens is quite complex, but its basic workings can be understood with some simple geometry. The method exploits the fact that what we see as images depends on the alignment of the object, lens, and observer. When the alignment is not quite perfect, the rays that form each of the images take different lengths. So one image appears to us before the other, ensuring that images from a gravity lens offer an automatic instant replay! Since the time difference implies a difference in the path lengths, it is possible to take the angles of the bending and calculate a geometric solution to the distances, much as surveying uses trigonometry to get distances.

Another use of lenses is as an interpretation of how many galaxies there are in the universe. Although the statistical arguments are rough, it is easy to see that the more gravity lenses you find, the larger the number of galaxies there must be; in other words, to increase the chance of seeing many lenses, one needs to increase the number of galaxies in the universe. Finding one or two gravity lenses may not say much at all about the number of galaxies (and how they are distributed), but finding several dozen does. This means that the gravity lenses can be thought of as a rough approach to a source counts test, although this approach has yet to be realized.

Finally, lenses yield a new way to understand the role of galaxy evolution, since some galaxies are actually cosmic telescopes, magnifying the view of very distant background galaxies. The magnification can be many magnitudes, so that galaxies that might otherwise be far too dim to see can be studied, if not for their structure then for their spectra. Perhaps very young, dim galaxies may emerge as lensed images by a far older, and closer, one.

These are the types of problems lenses help us with as we probe the galaxies and the universe. With a bit more understanding of the lensing process, it will be interesting to see what further insights gravity lenses have to offer, particularly through the view of the first lens, the Twin Quasar.

IMAGES FROM A GRAVITY LENS

Gravity lenses shape light, forming images that are magnified, inverted, reversed, doubled (sometimes tripled and more), and distorted, just like the images formed in a glass lens. The bending of light rays in a gravity lens follows many of the same rules as those governing refraction in a glass lens. This simple but compelling fact allows us to construct glass lenses that simulate the effects of gravity lenses. What we know about the optics of a glass lens can help predict what types of images a gravity lens might create.

Glass lenses designed to simulate gravity lenses have unusual shapes, and most look like throwaways from a lens factory or the bottoms of soda bottles. This points out a key difference—a glass lens is designed to magnify an object with great clarity and exactness, while nature's gravity lens is a lensmaker's nightmare. There are differences, however, in even such odd lenses, which correspond to the types of lensing that happen with a variety of mass and mass distributions for the bender.

This diagram shows what happens to the image of a point of light as a glass lens (designed to simulate a gravity lens) is placed at different positions, changing the alignment. Top view shows this magnifying glass at six different stages of alignment. Bottom shows side view with eye, alignment point (x), and point of light (.). Light gets deflected, and imaged, depending on the alignment; the Einstein ring occurs with perfect alignment.

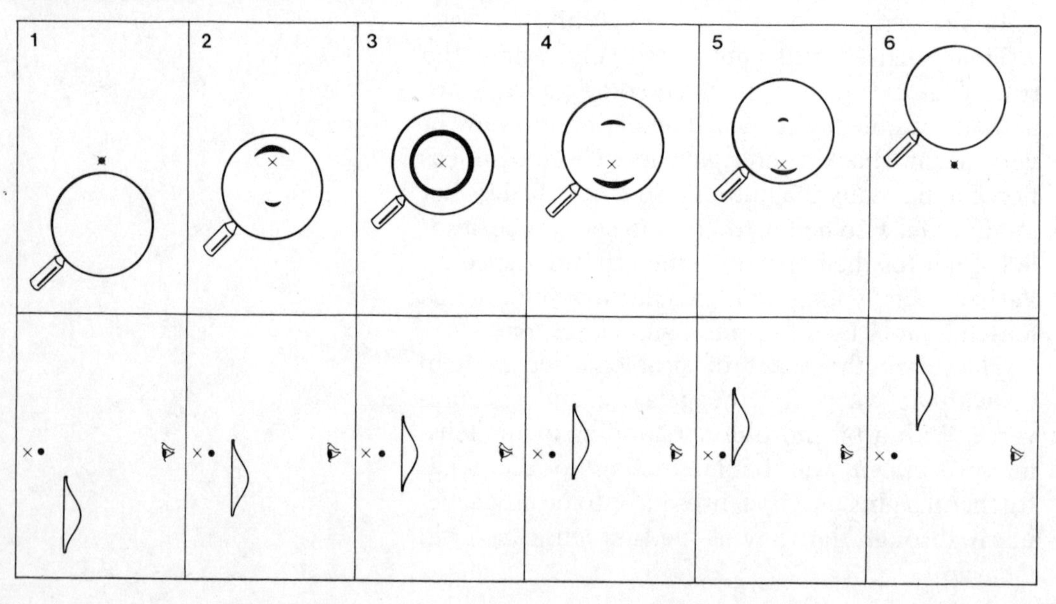

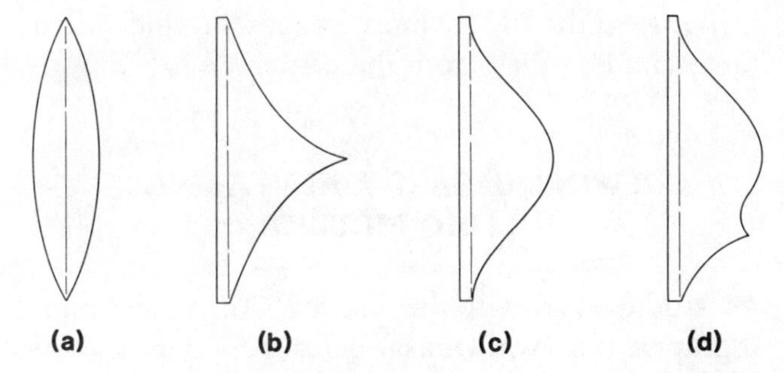

Side views of a variety of glass lens shapes that simulate gravity lenses: (a) is a normal lens, (b) simulates a black hole gravity lens, (c) and (d) simulate two galaxies as gravity lenses.

A black hole creates a different image than a galaxy would; a star cluster causes yet another effect. As a result, the glass lens counterparts to these differing mass distributions look quite different from one another. A black hole simulation, for example, has a pointed center, while the others are bulged in the middle in various ways.

Looking at objects through these looking glasses is a great way to simulate the making and breaking of alignment with the observer, the Einstein ring, and so on. With perfect alignment, the Einstein ring gives a tremendous degree of magnification. The ring's angular size depends on the mass and the distance of the bender. As alignment breaks, the crescents form, but the relative orientation and structure of the crescents depend on how much mass there is in the bender and how the mass is distributed and spread out. As the alignment lessens, the ratio of magnification for the two crescents disappears. Eventually the alignment becomes so skewed that one of the crescent images disappears. At this point, the observer sees a *caustic*—the transition point where the lens goes from making one crescent image to two.

Complex mass distributions, as one might see in a cluster of galaxies, complicate the view of the crescent images. In this case, other images, unrelated to the two crescents, form near the center of the Einstein ring. The shapes of the crescents may become irregular, no longer forming mirror images of one another, even with perfect alignment. Finally, the center of the mass distribution may still lie between the two or more images, but individual parts can be offset from the center.

TWIN QUASAR—A REVEALING RADIO PICTURE

As studied through the use of CCD's and photographs, the Twin Quasar gives a good indication of the relative magnification of crescents but does not reveal anything about their shapes; the crescents are too small to resolve with Earth-based optical telescopes. We need better views to probe these images. (See Plate 25 following page 142.)

When the Twin Quasar was finally studied with radio interferometry (see Chapter 4), the VLA radio pictures showed an ordinary radio source, a DRS, in which the sole jet was slightly curved. The core of the DRS was doubled (the second crescent), but the jet was seen only once; the core of the quasar must be positioned at a caustic.

The two core images coincided with their positions as visible light images. The core images are very small in angular size and do not reveal their detail with the VLA.

However, VLBI, with its of higher resolving power, has yielded much better details of the core images. Two VLBI studies in 1981 and 1983 showed that each core image had a jet emerging from it, although the fine details of these jets were not clear. But clearly the presence of this "mini-jet" in both core images made it possible to compare the images of a gravitational lens directly—with a bit more thought and observation.

In 1981 the French astronomer Claude Vanderriest made a perceptive prediction concerning the A and B core images. He surmised that the Twin Quasar may represent a DRS where the core is squirting out clouds of electrons at superlight speeds (see Chapter 4). If this was so, then each image would show electron clouds emerging, with differently magnified views of this phenomenon. It

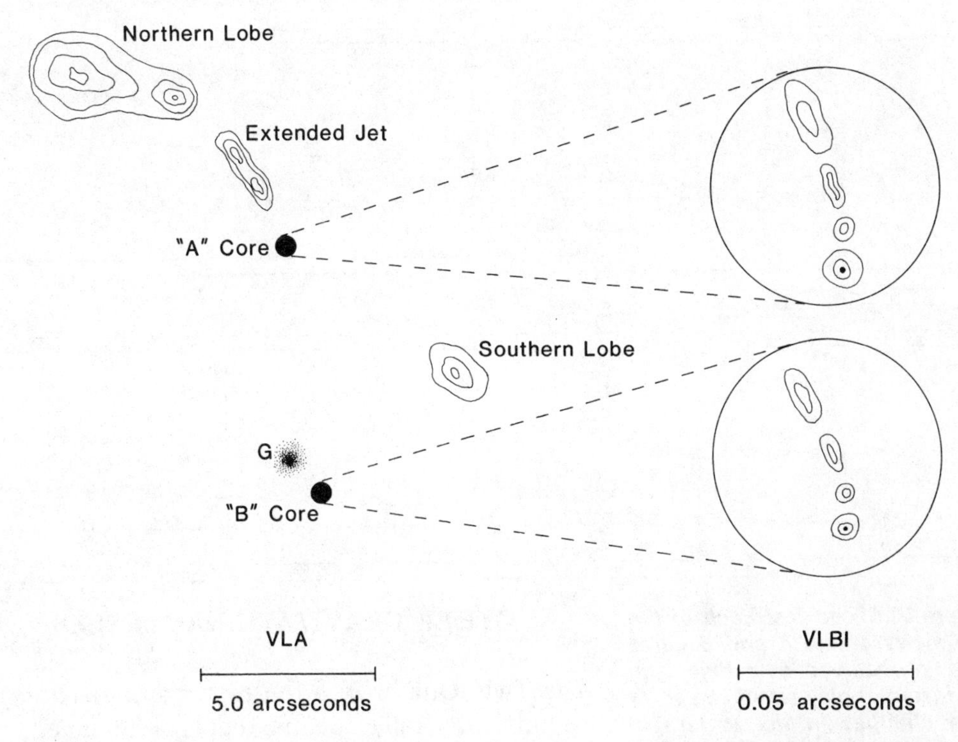

Radio views of the Twin Quasar. A and B represent the two images; labeled as cores here, they correspond to the positions of the visible-light images. The VLA view shows the extended jet and radio lobes for A; the B image only replicates the core. G is the position of a bender galaxy, one of many in the bender cluster. At right, superimposed views show the VLBI detail of the A and B cores.

should be possible to see the instant replay of a shot-out electron cloud when comparing A and B. And since the A and B core images take slightly different times to get to us, we can use this timing difference—a time delay—to find the distances of the quasar and bender. With the distance known, the Twin Quasar, through its redshift, acts as a calibrator for Hubble's constant. What Vanderriest envisioned was to use the magnification properties of A and B, along with the time delay, to reveal the geometry of the lens arrangement, the mass of the bender, and the distances.

Since that time, it has become clear that following the travels of superlight-speed clouds is a slow activity. Even though the cloud is moving very close to the speed of light, its change in angular position is incredibly minute—only a fraction of a thousandth of an arcsecond in one year. To measure an actual change requires years of observations, and VLBI movies are now being compiled to check Vanderriest's prediction.

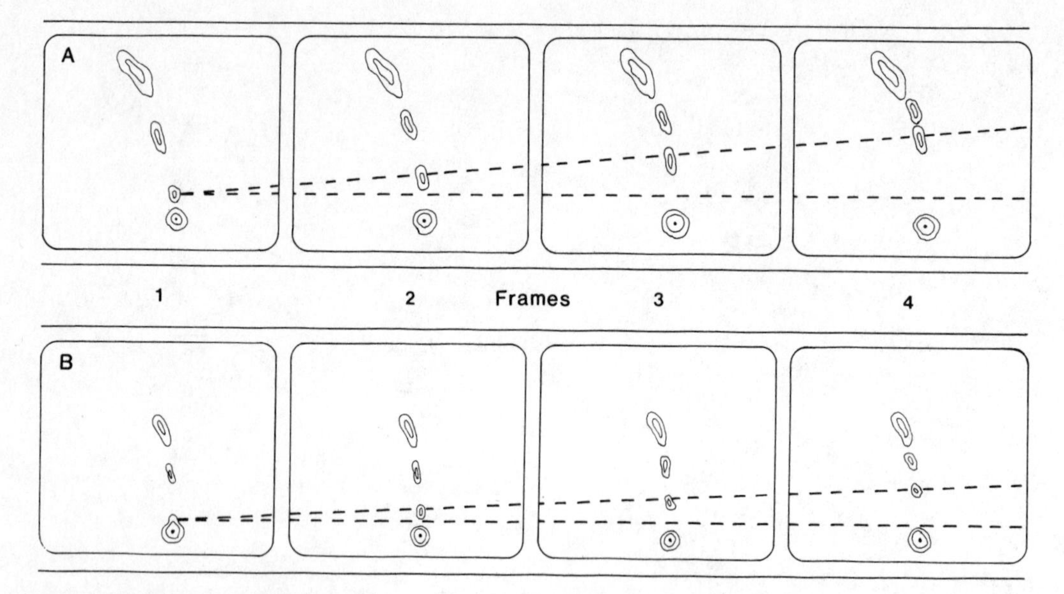

Future VLBI movies: Each of the two views of the A and B cores can be observed over time, and the instant replay effect can show how changes in the motion of clouds in the jet reveal the time delay, and thus the inferred distance of the bender and quasar. Hubble's constant can be found in this way. Each frame has a scale of 0.07 by 0.08 arcseconds.

OTHER GRAVITATIONAL LENSES

The Twin Quasar was the first gravity lens to be found, and it still is the one that is best studied. Yet over a dozen other gravity lenses have been discovered since 1979, and recent work reveals other ongoing insights. These gravity lenses can be divided into two general groups—those in which it is a quasar that is the mirage and those in which a background galaxy undergoes the gravitational lens effect. In these two groups, the bender is either one galaxy or a group of galaxies in a cluster. (See Plate 26 following page 142.)

In the gravity lens called 3C 324, a distant galaxy has been lensed by a nearby one. This makes for a curious situation, because previously it was thought that all we saw was the distant galaxy. The distant galaxy was anomalous because it was much brighter than one might expect from the Hubble diagram's interpretation of its redshift. This galaxy had been dubbed overluminous. In 1986 the French astronomer Olivier LeFevre revealed that the center of the visible galaxy of 3C 324 was at an entirely different redshift from that of the galaxy itself. Here we are seeing the perfect, or near-perfect, alignment of 3C 324 with a small, and much closer, galaxy. The two crescents are separated by a few arcseconds, suggesting that the nearby galaxy is quite massive, despite its apparent dimness. However, since it is very difficult to separate the close galaxy from the images (crescents) of the distant one, it is not possible to tell much about the bender itself or to look for an instant replay effect.

Another, similar case entails the lensing of a distant quasar by a relatively nearby, large spiral galaxy. At the core of the spiral galaxy, at least four bright images (an Einstein ring?) appear of a quasar with a much smaller redshift. This situation is striking because the galaxy's arms can be easily seen; it is as if a luminous ring hovers about the core of the spiral galaxy. This lens system, called 2237+031, was discovered in 1985 by the American astronomer John Huchra and his colleagues. It has a separation of only about an arcsecond for the images, but the placement of the images about the galaxy's core implies that a very massive point-mass (black hole) may lie at the heart of the core. When better views can be had of the images, it should be possible to figure out exactly how much mass in the spiral galaxy's core is pointlike, or how much gets attributed to dark matter.

Another gravity lens system may prove especially interesting. Discovered in 1987 by the Belgian astronomer Jean Surdej and his colleagues, it is named UM 673. It is a distant quasar being lensed by yet another galaxy bender. What is unusual is UM 673's brightness and variability. It was long suspected of being too bright even for a quasar. Indeed, the lensing may make this quasar several magnitudes brighter than if the lens were not present. This is actually an argument posed by Jeno Barnothy for quasars in the 1960s. Today, we must accept the reality that a fraction of the quasars we see may be "overbright" because of lensing.

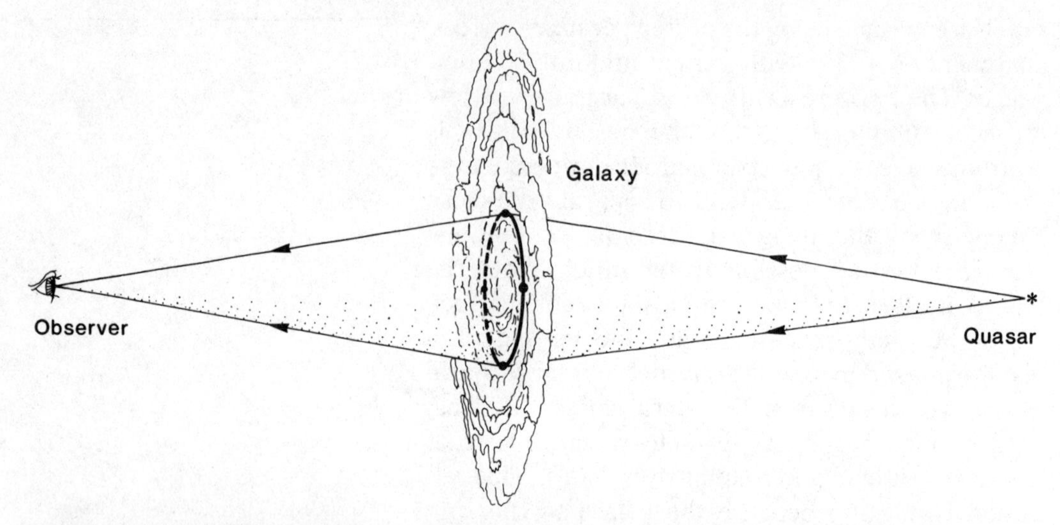

The gravity lens 2237+031 may be a case where the massive core of a spiral galaxy is aligned with a distant quasar, giving an Einstein ring.

Another factor in the interest of UM 673 is that it varies in brightness, just as the Twin Quasar does. However, the images of UM 673 lie closer together in angle, and the time delay for seeing an instant replay is short—months rather than years—so UM 673 might reveal its geometry and mass sooner than the Twin Quasar or any other gravitational lens. (See Plate 24 following page 142.)

Might we expect gravity lenses to give us a truly cosmic view of the universe? The young but emerging studies point out the great potential for detecting dark matter, finding masses and distances, and viewing young galaxies. We may expect several intriguing surprises from gravity lenses in the not too distant future.

· EIGHT ·

Swiss Cheese Universe—Bubbles, Voids, Streaming, and Clusters

AS we look out toward the visible edge of the universe, we enter a realm of great unknowns. For it is this part of the universe that, because of its faintness, had previously provided few clues of its adolescence. It is a place where cosmologists need to find out what the universe is like—to monitor the Big Bang, to find out how galaxies form, to see the effects of the universe's expansion. Recent studies of quasars and galaxies have indeed given some useful insight into these processes. But the biggest surprise about this faint universe was an entirely unexpected one; it comes from the large-scale grouping, or *clustering* of galaxies. Here we consider what it means to have galaxies form large clusters, and what it implies about the Big Bang and the universe.

THE CASE OF CLUSTERING

The realm of galaxies is a social one: Few galaxies are isolated from others; most form neighborhoods of many galaxies, called clusters. In Chapter 5, we saw that clustering leads to often antisocial circumstances, and galactic cannibalism and mergers are the price of galaxies forming groups.

Our own Milky Way is but one of more than a dozen galaxies joined in a small and loosely connected cluster called the Local Group. These member galaxies are various in form. Some are very tiny galaxies called dwarf galaxies; others are spirals, such as the Milky Way. Yet from the standpoint of the number of galaxies or their structures, the Local Group is nothing but

Virtually all the bright points of light on this photograph are galaxies, lying in a cluster in the constellation Perseus. (Courtesy National Optical Astronomy Observatory.)

an ordinary cluster. Elsewhere, to distances of many millions of parsecs, the clustering of galaxies can become spectacular, with hundreds of hundreds of galaxies in a cluster.

Early studies revealed that these giant clusters show little shape or definition, but there is generally one giant elliptical galaxy more massive and brilliant than the rest. This galaxy, usually at the rough center of the cluster, may be the cannibal galaxy, drawing others in and gobbling them up. The giant clusters are mini-universes in themselves because the gravitational pull of the collection controls the motions of all members. Astronomers loosely describe these clusters as *superclusters;* the relatively strong control such clusters exert on their own galaxies gives them a small autonomy from the universe as a whole.

The study of galactic clusters is a relatively modern one. Most astronomers have been concerned about clusters from the point of view of how galaxies evolve when in neighborhoods. Others were interested in the intergalactic medium, which often seems to be slightly denser in rich clusters (in which many members are closely spaced). But by the mid-1970s a number of astronomers were becoming interested in the size and comparison of clusters to see how they might connect to one another.

In 1977, the American physicist Jim Peebles and his colleagues noted that if one looked at galaxies as points on a map, connecting the dots showed something unusual. Instead of galaxies and clusters being spaced randomly, there seemed to be hints of a *large-scale structure,* a subtle, three-dimensional pattern in the way clusters were spaced relative to one another. But these hints were hard to follow up—each galaxy needed its spectra taken, and getting the spectra of thousands of galaxies, many of them quite dim, was not an easy task. It was like constructing a jigsaw puzzle with only a few of the key pieces.

Not too many years ago, astronomers could get an object's spectra only by running its light through

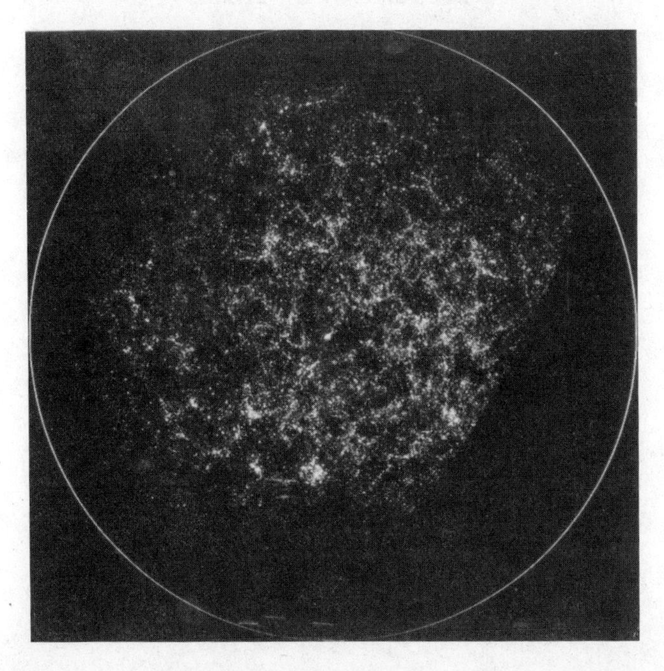

The million or so galaxies in this picture reveal the strange clustering that galaxies, and galaxy clusters, make on the large scale of the universe, here shown stretching across hundreds of megaparsecs. (From M. Seldner, B. L. Siebers, E. J. Groth and P. J. E. Peebles, **Astronomical Journal,** ***82, 249, 1977.)***

a prism. Today, however, it is possible to combine a finely etched grating and a prism to get a "grism." Grisms are one of the new techniques for getting spectra. Their main advantage is that every object in a telescope's field has its spectrum taken simultaneously, rather than singly. This makes it possible to obtain redshifts for many hundreds of galaxies at the same time and to map out the positions of many thousands of clusters, many at very great distances. Astronomers began to ask how clusters of galaxies at extreme distances compare with those that lie much closer to us. Many assumed that the answer was clear: The universe, to be isotropic, should really show no change between the shape and size of clusters at any particular time or region; galaxies should cluster the same way now as they did 10 billion years ago. Clusters may be a phenomenon of matter coming together, but the clusters should never become so big as to go against the general expansion of the Big Bang. Yet these new studies gave up an extraordinary result.

THE BIGGEST HOLE

In 1979 the American astronomer Robert Kirshner was studying the galaxies in the direction of the constellation Bootes. His equipment allowed him to look at very faint and distant galaxies and gave him the confidence that he could construct a picture of the way galaxies cluster over many slices of sky out to extreme distances. Indeed, Kirshner found many clusters of galaxies, some not previously documented. But he was startled to find that over a certain range of redshifts, and hence distances, no galaxies were seen at all. He could clearly see galaxies beyond this redshift range but not within it. Effectively, Kirshner had found a place where galaxies—single or in clusters—were virtually absent.

Here was an odd phenomenon—a *void,* or hole,

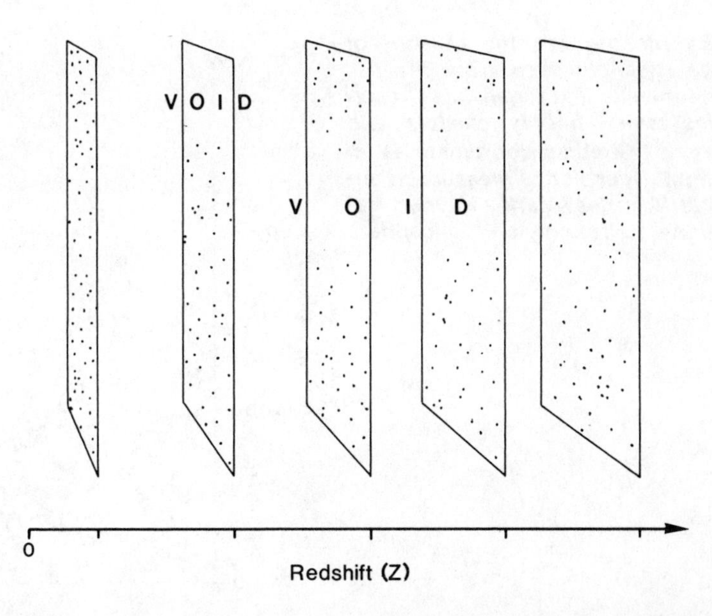

A slice of sky—looking at a patch of sky and dividing it into different redshift slices.

in the universe. At first, some astronomers doubted the reality of the hole; no other studies of galaxies had revealed anything like this Bootes void, which was many billions of kilometers in breadth. It seemed hard to believe that a void could exist; the small motion of galaxies and clusters tend to fill in any holes that might arise in the universe. Yet independent study showed that Kirshner's void was real, not merely the result of incomplete observation or interpretation.

By 1985, many other cosmic voids had been found in other regions of space. Some were a fraction of the size of the Bootes void, while others were dozens of times larger. The cosmic voids, hints of which showed up in Peeble's work, were a whole new and commonplace property of the universe.

A curious aspect of these voids was the number of galaxies clustered at their edges. Further study revealed that the clusters of galaxies bordered the voids in a way that showed a gigantic filamentary structure. In a poetic mood, some astronomers called these filaments the *Five Fingers of God,* after their handlike structure. The Five Fingers indicated

By plotting out the redshift of galaxies observed from the 21-centimeter hydrogen line, galaxies reveal bubbly clusters and voids. Circular coordinate is in right ascension, a measure of angle. (Courtesy Martha Haynes, National Astronomy and Ionosphere Center.)

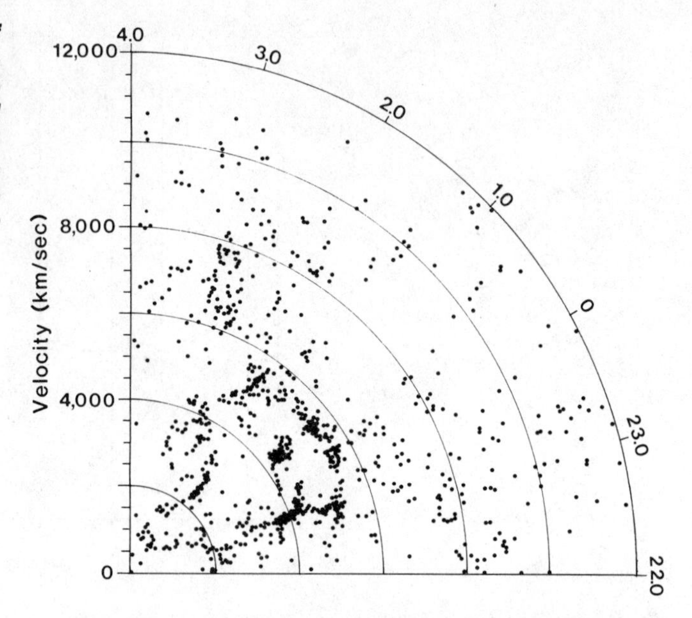

a tendency for galaxy clusters to collect along lines of thousands of billions of parsecs, lines broken at distances relatively near to us, where the galaxies in the aligned clusters had moved just enough to fill in or render insignificant any nearby voids.

According to the distances calculated, these great voids and their surrounding structured groups of galaxy clusters seemed to be common in the universe until very recent times. Just after the formation of galaxies, the universe was not a smooth and random distribution of galaxies but a highly ordered one, extending to times billions of years ago.

Do voids and filaments make sense with a Big Bang model? In the strictest sense, they do not. Galaxies, and thus their clusters, are the remnants of an isotropic explosion, in which the velocity of the expansion causes galaxies at different distances to move with different velocities according to Hubble's law. We speak of galaxies as moving with the *Hubble flow,* the expansion velocity of the universe. This flow should prevent galaxies and clusters from connecting over very great distances.

COSMIC BUBBLES AND STREAMING

The early studies of the voids and filament structures were limited by the sensitivity of the equipment used, providing just a hint of the actual large-scale structure of galaxy clusters. By 1986, astronomers were finding that various slices of sky showed more than a mere void/filament structure to galactic clusters. Rather, the voids resembled bubbles, with the galaxy clusters surrounding them forming a sort of sponge; the overall pattern of the cosmos looked something like Swiss cheese rather than filaments. The evidence for this spongy or Swiss cheese structure is compelling and may best be shown in astronomer Brent Tully's 1987 map of the galaxy clusters. He has plotted the redshifts and positions for thousands of clusters and presented them as a three-dimensional map. Stretching across many billions of miles of space are long tubes of connected galaxy clusters separated by vast voids. (See Plate 22 following page 142.)

While some astronomers were exploring the bubble/void structure, others were trying to solve a troubling puzzle presented by the galaxy redshifts. The Big Bang model describes the expansion of the universe as a universal effect—no part of the cosmos is exempt. Yet galaxies aren't perfect markers for the pattern of expansion; they show evidence of a slight motion that is caused not by cosmic expansion but by the slight gravitational pull of neighboring galaxies. In general, this so-called random motion is smaller than the effects of expansion, and all galaxies can still be described according to the Hubble flow of the expansion.

But some galaxy clusters show a difference in redshift when compared with others at apparently the same distance. When these distances are estimated using the Tully–Fisher relation (see Chapter 4) and other magnitude methods, the clusters seem to be moving either faster or slower than the recession velocity that the Hubble law suggests for their

The Hubble flow and random galaxy motions. Galaxies move with the redshift that corresponds to their distance, but there are additional velocities caused by their random motion.

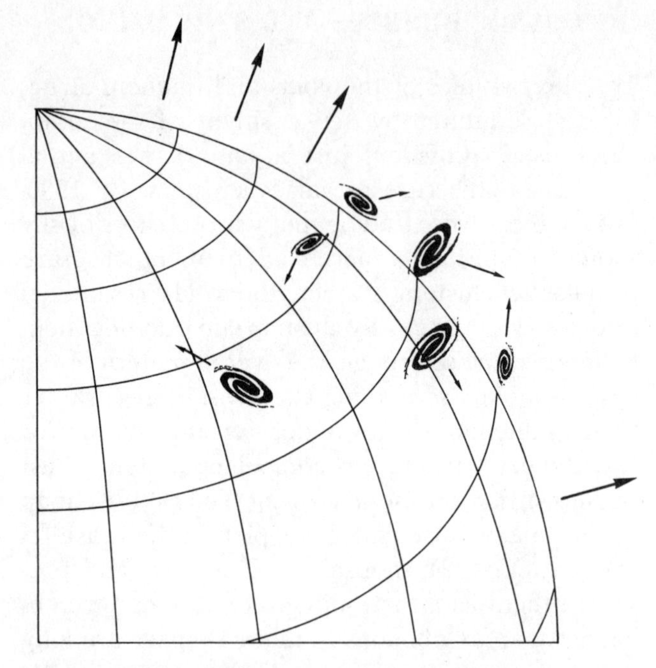

distance. The difference may be as much as 10% to 20% of the velocity expected from the expansion of the universe at their distances. So to a marked degree the expansion of the universe is *not* totally isotropic but changes from one cluster of galaxies to another. This large departure from the velocity of the universe's expansion is called a *streaming* motion and is much larger than the effects of the small random motions seen in clusters.

Hints that streaming existed actually came about from the detection of anisotropy (see Chapter 4) seen in 1978. This slight Doppler shift indicated a slight motion of our galaxy relative to the CBR (and thus the Hubble flow). This motion has a random component in it, but part of it may be streaming as well. Even earlier studies in the early 1970s, by astronomer Vera Rubin and her colleagues, showed that galaxies in a large cluster of galaxies in Virgo were moving at speeds, and a direction, not accounted for by the Hubble flow. By 1985, many more cases of the streaming motion were seen.

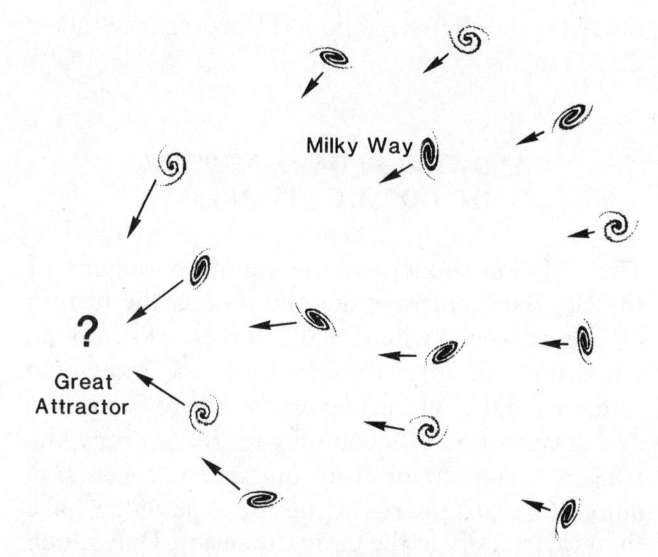

Streaming—the motion of galaxies that indicates a convergence toward a central point. This motion far exceeds random motions and can be greater than 20% of the velocity that a galaxy has as part of the Hubble flow. Is it caused by a Great Attractor?

As of 1987, the streaming of galaxy clusters appears to be a common phenomenon, especially among clusters that are over 100 million parsecs from Earth. Along with other galaxies in the Local Group, our galaxy is heading toward a central point in the direction of the constellation Virgo. The mystery of why many galaxies and galaxy clusters seem to be converging toward a central point might well be the result of a strong gravitational pull. Astronomers speak of the *Great Attractor*—a huge, invisible, unknown mass—which is pulling our galaxy toward it, along with hundreds of other galaxies. But we cannot detect any luminous object at the position where the Great Attractor appears to be. How could it be so influential yet remain invisible?

Streaming establishes that the galaxies in clusters don't move the way a simple model of the universe's expansion might suggest. The expansion, in general, is not debated, but the streaming adds an intriguing point—what is its cause? Is something, perhaps unseen, pulling at clusters, lining them up, and forcing them to move? What could cause streaming in the context of the expanding universe—or is the universe not the result of a Big Bang? In failing to account for this motion, how flawed is the Big Bang model? What crucial factor does it lack?

ANSWER(S): DARK MATTER OR COSMIC STRINGS?

The inflation model for the earliest moments of the Big Bang portrays the one time in the history of the universe when all its forces were on an equal footing. Physicist Alan Guth was motivated to describe the infant universe's radical expansion as a cornerstone in accounting for the isotropy; the CBR is a spin-off of this time, and it appears so smooth because parts of the universe didn't have time to diversify in the early expansion. Only much later on, as matter cooled to form galaxies, did gravity pull together clumps that became galaxies.

But as pointed out earlier, galaxies could hardly have formed without sources of mass that seeded the clumps—seeds that may have been around from the beginning of the universe. What created these seeds in the first place?

One of the answers might lie with the cosmic strings. The seeds may be the result of defects from the symmetry breaking at the Big Bang's beginning. Frozen, like air in ice, these defects represent bizarre bits of space containing very high energies.

Cosmic strings are different from ordinary matter. Although they are not mass, the strings are distortions in space and time, so they produce an effect of great mass concentration. To us, they would appear as incredibly massive, yet narrow, filaments, only an atom or two thick but up to many thousands of megaparsecs long. They have the same gravitational pull as a cluster of galaxies wound up into an incredibly narrow filament and stretched out over countless billions of kilometers in space.

Some cosmic strings might be the perfect seed for forming galaxies, since their strong gravitational fields act as natural attractors to all matter. The theories behind the hypothetical cosmic strings show that they have two curious properties: They form loops (no string has ends), and they "evaporate," dissipating over time and fizzling away in bursts of energy. Only a small fraction of any possible strings could remain the 10 billion or more years since the Big Bang. But in the early universe, there may have been many more seeds than now.

If cosmic strings exist, we might expect them to make the CBR slightly less than perfectly smooth from point to point. Instead, these strings would cause small variations in the makeup of the universe. The CBR should show a very slight "grainy" anisotropy from the cosmic string seeds. As such, the main thrust of the need for isotropy in a Big Bang is not contradicted but modified. These mass magnets were made by the Big Bang; only after the early expansion do they start showing their effects. And these effects are profound. The cosmic string can draw in galaxies, forming a void of captured galaxies and a bubble of enhanced galaxies. It can cause galaxies to stream through this attraction. Indeed, the Great Attractor may well be an invisible cosmic string.

Yet cosmic strings might not be the only way to get the Swiss cheese universe. In 1984, American astronomers Joan Centrella and Adrian Melott undertook computer simulations for the growth of clusters of galaxies where there was much dark matter present. (See Plate 21 following page 142.) They showed that the Swiss cheese structure of the universe makes sense if the dark matter was around to guide the growth of galaxies, forming the large-scale structures, bubbles and voids. What was this dark matter made of? Was it composed of the mysterious neutrinos or of something else?

In any case, there was a need to find out more about how this seeding takes place. This need was to be addressed by yet more sensitive observations of the CBR.

THE NOT-SO-PERFECT CBR ISOTROPY

The CBR showed a slight change in intensity over different sky positions, an effect that was interpreted as indicating a Doppler effect motion of Earth. But there is another, smaller change in the CBR that suggests seeds—cosmic strings or some form of dark matter—were around during the time of the CBR.

The observation, made by the British astronomer Richard Davies and his colleagues in 1987, showed that the CBR fluctuates in intensity from one part of the sky to another by a ratio of 40 parts per million. That is a very small change and one that does not have the directional effect that one should see with a Doppler effect. The fluctuations may represent the imperfections—seeds—that can later lead to the growth of galaxy clusters. They are only beginning to act as mass magnets, pulling together and forming clumps of hydrogen to make some regions slightly brighter than others. Cosmic strings are consistent with this observation, although they may not be the only realistic explanation for it. Yet if the CBR fluctuation indicates strings, what would the effects of the strings look like later on in the universe's growth?

A COMPLEX BIG BANG

Cosmic strings are very appealing as an explanation for bubbles and voids, galaxy formation, and so on. But their influence does pose complications in performing cosmological tests. For example, if the strings add streaming to galaxies, especially very distant ones, then the galaxies are not perfect markers of the Big Bang's expansion. It may not be possible to separate the effects of streaming and galactic evolution from those caused by the universe's expansion and deceleration. And it is still not known if the quasars, the most distant of galaxies, form bubbles and voids as well as streaming.

Until these aspects of the cosmos' large-scale structure are fully understood, clever ways must be devised to find the cosmological parameters and to see if the Big Bang can be modified to account for this structure. And if the Swiss cheese universe is mostly a type of dark matter, what can it be? Or, perhaps more accurately, what can it not be?

· NINE ·

Dark Matter—What and Where Is It?

EVERYTHING we know about the universe ultimately comes from what our instruments and eyes tell us. With the exception of what gravity lenses reveal, the only universe we can know much about is the universe of radiant matter, even if it constitutes only a fraction of the full picture.

But we cannot ignore the invisible universe. Put in another way, the dark matter problem confronts us in many forms, including such questions as whether the universe is open or closed, how galaxies stay together as distinct entities, and how the clusters of galaxies maintain their groupings over billions of years. These fundamental questions have prompted astronomers to consider the problem with great intellectual urgency, so that we might have some notions of what the universe we can't see is like—if not now, then in the near future.

LITTLE PROBLEMS, BIG PROBLEMS

Looking for dark matter is like looking for black boulders at night while wearing dark sunglasses. You may be surrounded by boulders, but their presence remains elusive. Perhaps you might suspect their presence because the boulders funnel the breeze around them or because rain splashes off their surface, dripping into puddles. You have used indirect observations to conclude that they must exist, even if you can't see them. Now place those boulders in space, perhaps a few parsecs away, perhaps a megaparsec away. Make the boulders bigger than Jupiter or pulverize them into a swarm of elementary particles.

Either way, the boulders remain; there could be lots of them, and you can only guess that they must exist by their indirect effects on other things.

The dark matter problem is the more generalized idea of cosmic boulders. The evidence for dark matter is indirect, based almost entirely on the conflict of how galaxies move or stay together. This conflict compares observation with theory, pointing out how we must be wrong about our understanding of gravity (a farfetched explanation) or that much of the matter is invisible—cosmic boulders of a plethora of potential forms.

The closest example of this conflict entails our nearby region of the Milky Way. All the stars, including the sun, orbit around the center, or core, of our galaxy, taking over 100 million years to complete a single revolution. While the revolution may naively be viewed as pulling the matter away from the core, it is counterbalanced by the gravitational tug of each individual star on all the others, particularly the dense and massive region at the core. Like the planets orbiting the sun, stars near the core orbit relatively rapidly (less than a 100 million years), while those much farther out orbit many times more slowly.

By looking at many different stars at different galactic positions, it is possible to plot the velocity of the stars' motion compared with their distance from the core. This *rotation curve* is a way of testing this comparison between how the stars orbit, as estimated by theory, and their actual motion, as observed. Taking our galaxy as a pinwheel-shaped one with a massive core, the theory shows that the stars close to the core move very fast and that those much farther away from the core move only slightly slower! This contradicts a commonsense notion of the idea—called Keplerian motion—that the most-distant orbiting stars should be moving quite slowly.

The simplest way to explain this is to assume that the mass of the galaxy is not simply a massive

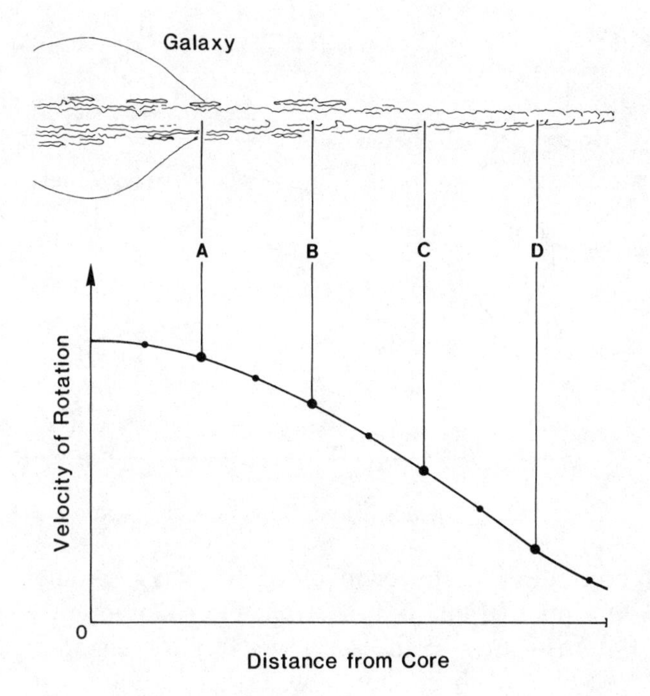

The rotation curve of a galaxy is a way of describing its mass through its motion at various points along its structure. Here the drop off at the center has been suppressed.

core and a lighter, rotating disk. Photographs may imply this to be the case, but models show that the unexpected rotation curve can be explained by a situation where the galaxy's mass is mostly outside the core and disk and some extends well beyond the disk (pinwheel) of the galaxy; much of it must lie at right angles to the plane of the disk itself. The mystery matter explains the rotation curve but doesn't show up in photographs. This dark, mysterious matter cannot be regular stars, because we can easily see such stars at relatively great distances. It must be some other type of mass that does not give off much light.

Many other galaxies exhibit this same behavior on an even grander scale. The best way this has been shown is by using radio astronomy to look at the rotation curve for the cold hydrogen clouds in these galaxies, through observation of the 21-centimeter line.

The rotation curves of the 21-centimeter spectral line show not only that the hydrogen ex-

The rotation curve of a galaxy. The top line shows the observed curve, while the halo and disk represent the separate contributions of a visible galactic disk and an invisible—dark matter—halo. Only both combined can explain what we actually see. (From "What's the Matter in Spiral Galaxies?" Vera Rubin, in Highlights of Modern Astrophysics, *Shapiro and Teukolsky, eds., Wiley, 1986.)*

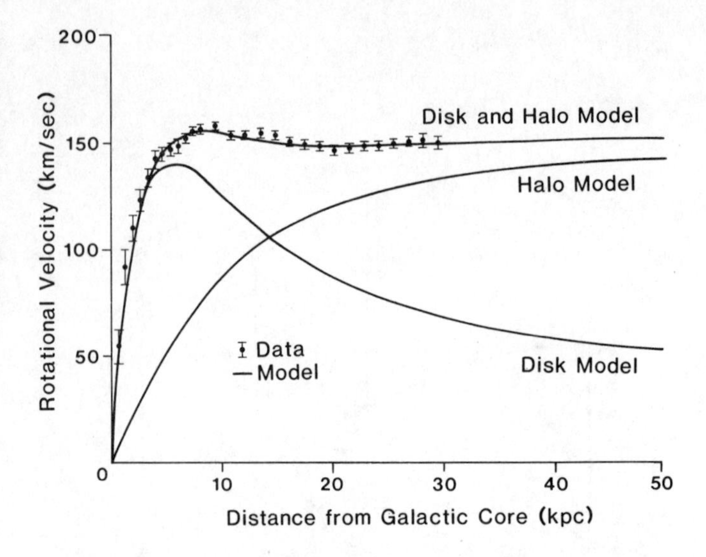

tends beyond the edge of each galaxy, as seen on a photograph, but also that the rotation curve flattens—stays at the same velocity for distances well beyond those that show up on a photograph. Based on calculations, the gas moving at the outer edges of these galaxies is moving much too fast for the core's gravitational pull to keep it within the galaxy—unless there's more matter than we can detect. No observed structure of stars can explain how this gas moves so fast yet stays in the galaxy. Again, there must be dark matter present.

Another example of the dark matter problem comes from studying the galaxies in clusters, the original (if not the only) way in which masses of galaxies could be estimated. Using the virial theorem, as discussed in Chapter 3, astronomers found the galaxies to have more mass than was guessed from the light being given off; their mass-to-light ratio, or M/L, was sometimes higher than 200-300 (the sun is defined as having an M/L of 1), which implies the presence of vast amounts of matter not contributing to the overall light of the galaxy.

Yet another puzzling question about galaxies is why they have the shapes they do. Using a computer, it is possible to simulate the formation of galaxies and to see how their shapes change after they have formed. It is difficult to simulate the shape of a spiral galaxy without placing a hypothetical cocoon of matter around the galaxy, a buffer that enables the pinwheel-shaped disk to form and retain its shape.

A particularly interesting example of this modeling problem is the *polar ring galaxies.* These are rare galaxies whose structure looks like a sphere of stars within a large ring of others. They may be related to the more conventional spiral galaxies, save for this strange ring effect.

These polar ring galaxies are not examples of gravitational lenses, although the confusion seems obvious. The redshift of the ring is identical to that of the star sphere, unlike a lens, where the ring would indicate an image of a foreground galaxy with much larger redshift.

The rings are real structures—yet they are impossible to explain by the mass of the sphere of stars. Rings are notoriously unstable and, like smoke rings, dissipate over time if left uncontrolled.

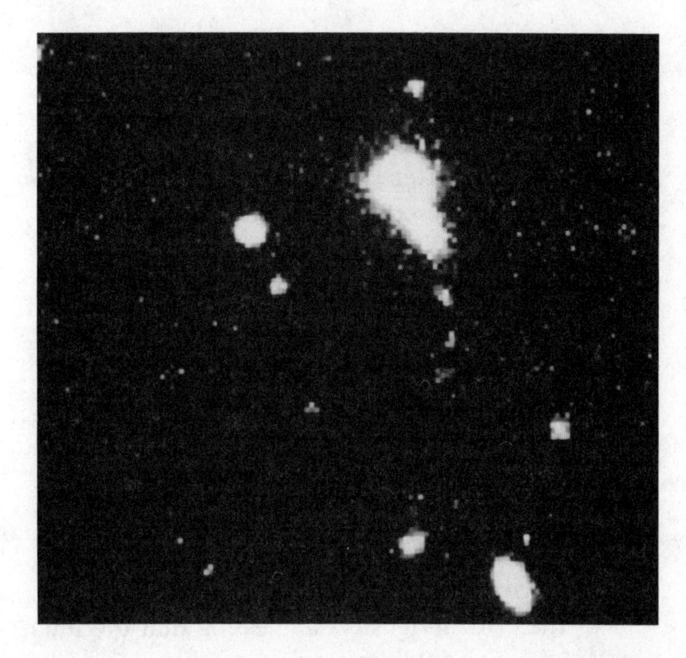

A very dim galactic cluster. Each bit of fuzz represents a separate galaxy of unknown type and mass. (Courtesy R. Schild, SAO.)

A spiral galaxy may look to us as in (a); in reality it may be surrounded by an invisible cocoon of dark matter (b).

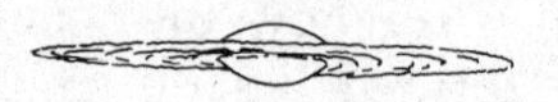

(a)

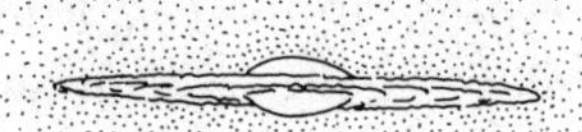

(b)

However, a ring can stay around for many tens of millions of years, if not more, if it is balanced by the gravity of matter lying beyond it. Indeed, several attempts to model the polar ring galaxies by extended, dark matter cocoons have had great success in showing that the rings can exist for long periods of time. The existence of polar ring galaxies demands the existence of an invisible, dark matter cocoon.

COSMOLOGICAL DARK MATTER

Recall that conventional estimates of the universe's density show that it amounts to only as much as 15% of the critical density, suggesting that the universe is open and will expand forever.

But recall, too, that the inflation model requires that the density of the universe equal the critical density, suggesting that the universe is just barely closed. And this model is a convincing one, for it describes why the universe exploded and why the CBR is so relatively smooth. The inflation model makes so much sense that many astronomers doubt that the estimated mass density is anywhere near correct. Rather than accepting the conventional observations and their conclusions, these astronomers are willing to admit that there may be a great deal of dark matter in the universe—85% of its mass! But 85% of the universe must be hard to hide; unless new methods and results show that the dark matter exists in such great quantities, the inflation model will have to be reconsidered.

There is another reason, albeit an indirect one, that so much dark matter makes sense. Theories of cosmic voids and superclusters require that massive "somethings" exist, perhaps made of matter that did not congeal as galaxies and thus does not burn brightly enough to see. Voids may not be empty; their contents are probably invisible to us with present observing methods and equipment.

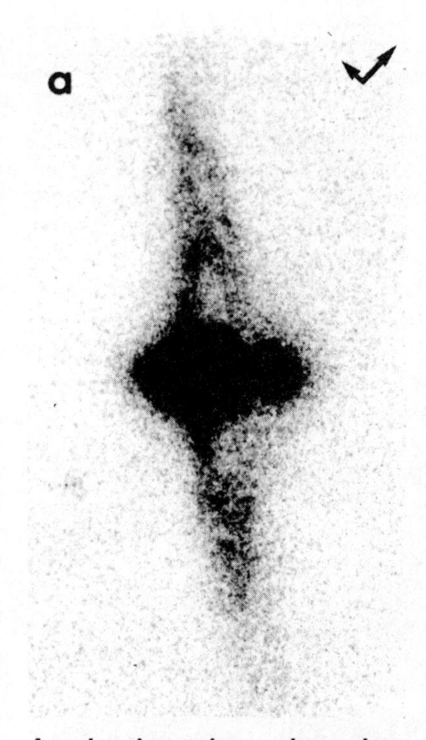

A polar ring galaxy, whose ring must be stabilized by a cocoon of dark matter. (From a paper published in the July 1983* Astronomical Journal *by F. Schweizer, B. C. Whitmore, and V. C. Rubin. © copyright 1983 by the American Astronomical Society.)

WHAT CAN IT BE?—DARK MATTER'S MANY FORMS

It may be easy to guess at the nature of the dark matter. Indeed, astronomers have made dozens of suggestions. Let's start by understanding what dark matter cannot be.

Dark matter, at least that which exists in galaxies, cannot be ordinary stars. Today's CCD methods and large telescopes are so sensitive to light that even the faintest of ordinary stars can be seen in many galaxies. If the dark matter is made up of stars, these stars are hundreds of times dimmer than ordinary stars. And in some galaxies, the rotation curves indicate that the dark matter may be as common or more common than ordinary matter—so any very dim stars must be more common than ordinary stars. Might we not see one or more of these feeble stars close by and scrutinize its properties?

Dark matter is not the hot gas that makes up the DRS's, nor is it the cold hydrogen we see from 21-centimeter-spectral-line studies. The mass of all this gas amounts to only a small fraction of the mass seen in the stars of the parent galaxy. This gas may be a good marker for the effects of dark matter, but it is not the dark matter itself.

Dark matter is not made up of stars that didn't make it—planets of large size. Such giant Jupiters might not give off visible light, but they would be strong emitters of infrared radiation, which would have been detected if present beyond the outer edges of some nearby galaxies. Yet several studies of this kind have failed to show any infrared radiation from very cool stars beyond the perimeters of the main structure of galaxies.

Then what is the dark matter? One viable suggestion is that it takes the form of strange stars—exotic variations of ordinary stars. In particular, two types of stars have been considered as likely candidates for dark matter: brown dwarfs and black holes.

Dark matter candidates and their individual sizes.

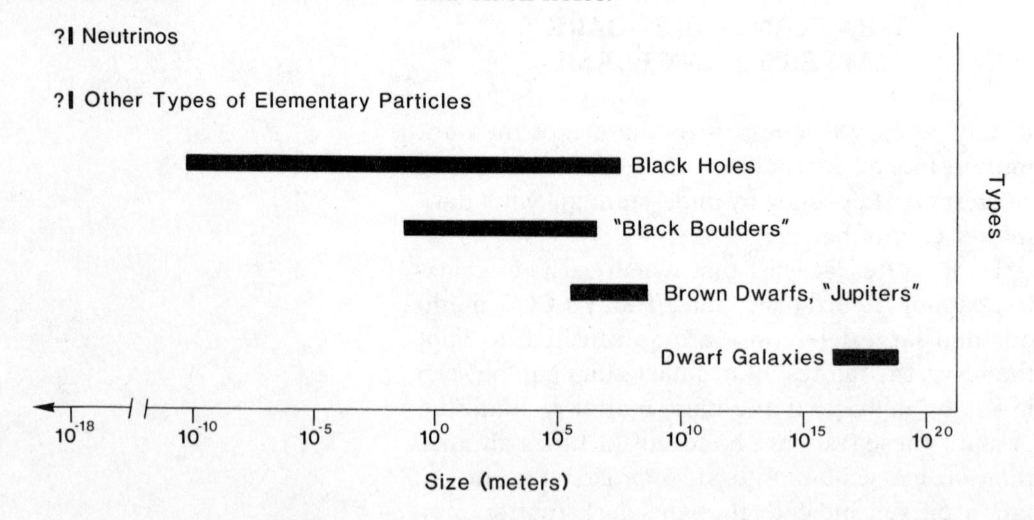

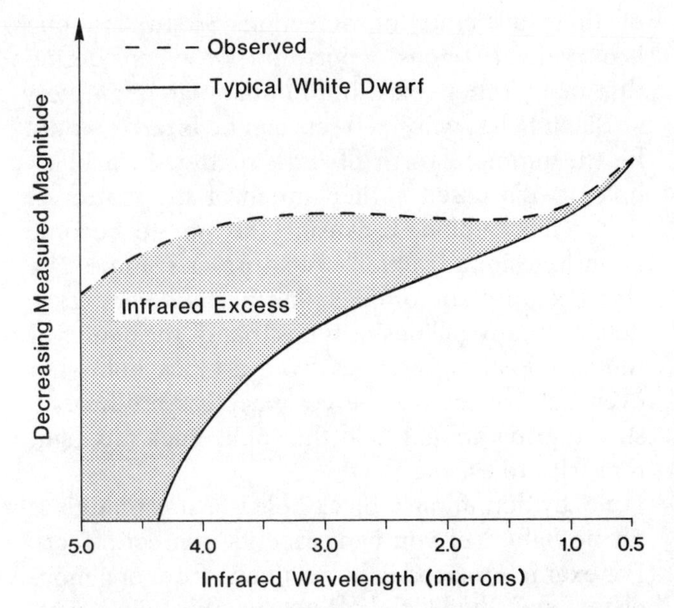

A brown dwarf can be found by seeing how much excess infrared radiation (from the brown dwarf companion) appears when looking at white dwarfs, which normally have very little infrared radiation. Spectroscopic studies are the best way (so far) of finding brown dwarfs.

Until recently, the brown dwarf was an unconfirmed type of star. A brown dwarf is a star with about 20% the mass of the sun, but which emits heat at about one-fifth the sun's temperature. This indicates that the very heart of these marginal stars is not quite hot enough to start up their hydrogen fires. Their radiation is all in the infrared. They may, in fact, resemble giant Jupiters more than stars like the sun.

The 1987 discovery of a brown dwarf (in our galaxy) came about after years of effort and a bit of luck. American astronomers Benjamin Zuckerman and Ed Becklin discovered that a white dwarf star, called G29-38, was emitting too much radiation in the infrared; the most likely explanation was that a much cooler brown dwarf was a companion to the white dwarf and emitted most of its feeble radiation in the infrared. The two stars are too close to each other to separate them with an infrared image, but perhaps they may be distinguished with future work.

The fact that no more brown dwarfs have been found is not an indication of their scarcity but of their difficulty of detection. So finding one brown dwarf opens opportunities for considering that many others exist but in unknown quantities.

Black holes, whose effects can be bizarre, would be the ultimate form of dark matter. In a black hole, as discussed earlier, much of the matter in a star has imploded, causing the star to become an infinitesimal fraction of its former volume. The star becomes so compact that it has a very large density, many billions of times that of the center of the sun. Nothing escapes from a black hole—not even light—because the compactness produces a strong gravitational field that pulls back any light that tries to escape.

So, by definition, a black hole is a star that gives off no light. Yet you can trace its indirect effects. For example, placed in projection in front of a more distant star, the black hole acts like Einstein's original gravity lens, forming a bright ring image of the distant star's grazing, but not trapped, rays. If other stars are companions to a black hole, the black hole will tear off their gas, forming a pancake-shaped accretion disk that can itself give off light. This pancake will move and vary, so it will not resemble an Einstein ring (which should also be much bigger). The brilliant accretion disk may be the indicator of the invisible companion.

Black holes may be present in another form as well. In 1976, British physicist Stephen Hawking showed that black holes could have been very common in the beginning of the universe and that once formed they might exist for billions of years. These black holes would have small masses; they would be mini black holes that could exist in vast numbers over the universe but foil us in our attempts to spot their effects. These mini black holes would be too light to act as gravity lenses and might not be common around other stars. They could easily remain in the outer hub of galaxies, where they would contribute to the galaxies' mass but not their light.

The smallest of the dark matter candidates are the elementary particles. These fundamental bits of matter were the prime form of the universe at its earliest moments, during the inflation period. Most of the particles lasted for very short periods of time—billionths of a second—but others may still be around today.

Much attention has been focused on neutrinos, those tiny elementary particles that almost never come into contact with matter. The Big Bang was such a hot and dense environment in its beginning that there could be countless numbers of neutrinos still around, contributing their tiny masses (if they have masses) as an aggregate amount. The universe, even today, could be a sea of neutrinos, their mass composing the majority of the universe's mass.

Finally, it has been suggested that tiny galaxies could have formed inside the cosmic voids. These galaxies are too dim for us to see, but their overall mass might be large enough to enhance the forming of other, visible galaxies that congregate as the bubbles.

Dark matter must exist, but its form is unknown. Let us now explore what recent observations say about the extent, if not the nature, of this mystery substance.

SENSING THE DARK MATTER

Some astronomers quip that dark matter is the result of insensitive photography, not matter in a mysterious form. And to a certain degree, they are correct. (See Plate 13 following page 78.)

In our studies of a few galaxies, extremely light-sensitive CCD images have been taken that show that the dark matter need not be invisible. Hypothetical cocoons of matter actually show up as diffuse and dim auras, filaments around these galaxies.

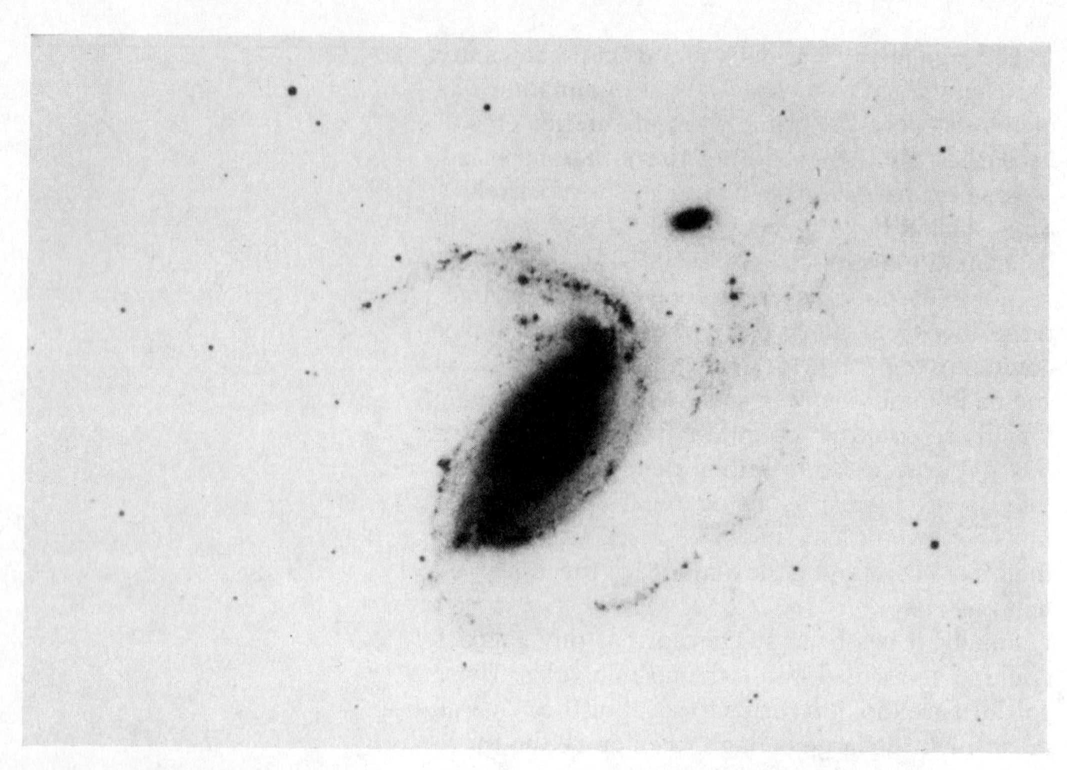

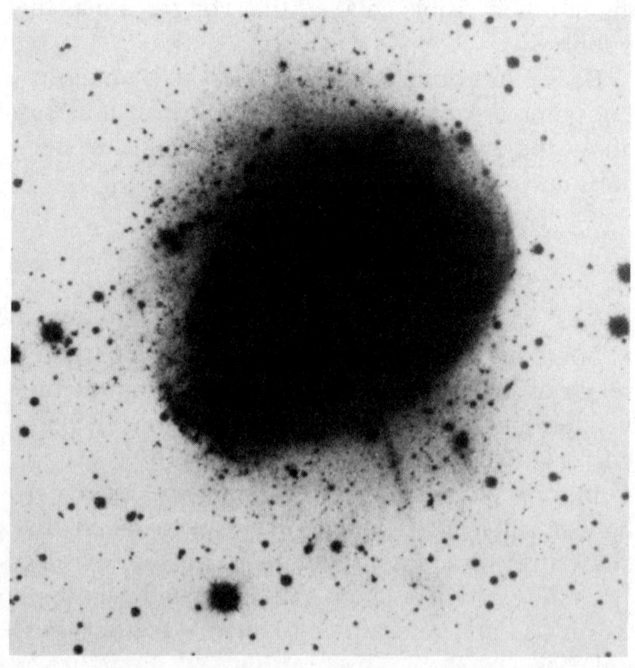

A barred spiral galaxy (above), showing its core and arms. The same galaxy (right), revealing the dim outer stars that may make up part of its cocoon. (Courtesy H. C. Arp.)

For at least one galaxy, the cocoon appears to be more than a few times the galaxy's previously observed size. In 1986, Canadian astronomer Sidney Van den Bergh showed that the giant elliptical galaxy M87 is surrounded by a haze of dim filaments, whose light is less than one-thousandth that of a normal galaxy. It is not known yet what fraction of this light arises from stars, rather than from hot gas.

These filaments are dark matter that give this galaxy its large mass. But the filaments themselves have a very large ratio of M/L. Is this a situation where some normal stars are mixed in with brown dwarfs or with other dark matter candidates? This unanswered question may be addressed by further observations in the future.

Perhaps the most important result of these findings is that the M87 cocoon is not merely mythical but has been clearly observed. How much of this cocoon extends to the neighboring galaxies of M87,

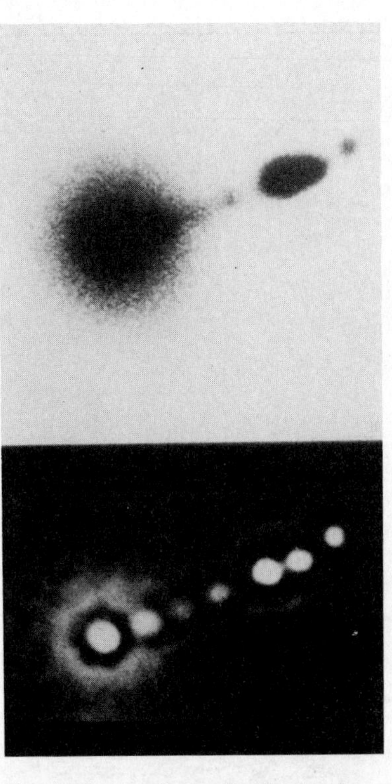

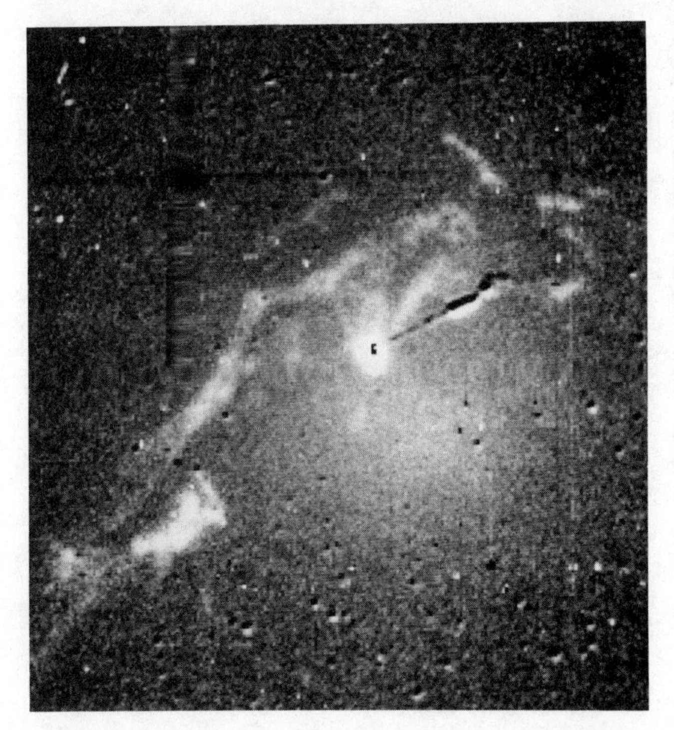

Many views of an elliptical galaxy called M87. Upper photo reveals the galaxy and a view of its visible-light jet. (Courtesy H. C. Arp.) Photo at left shows the same galaxy and jet (black) with wispy filaments surrounding it as a part of its lacy cocoon. (Courtesy S. Van den Bergh.)

and how does it control their shapes in addition to that of M87?

As suggested earlier, gravity lenses are another great way of seeking out dark matter. This is because the bender's mass is the principal factor in the shaping of the images of the more-distant background galaxy. At least half of the gravity lenses already seen do not seem to have benders that have been detected. In others, the bender galaxy or galaxies is present, but the images require the presence of more matter than the galaxy bender's light implies there is. This may be discrepant by a factor of ten or more; a particularly large discrepancy, say of 100 or more, would be the most direct indication of dark matter's presence, if not its nature. (See Plate 27 following page 142.)

Dark matter may be not only on the outside of

Supermassive black holes, the engines of DRS's, may have shown up in rotation curves of the inner cores of some spiral galaxies. The cusp in this idealized rotation curve corresponds to a large, compact mass.

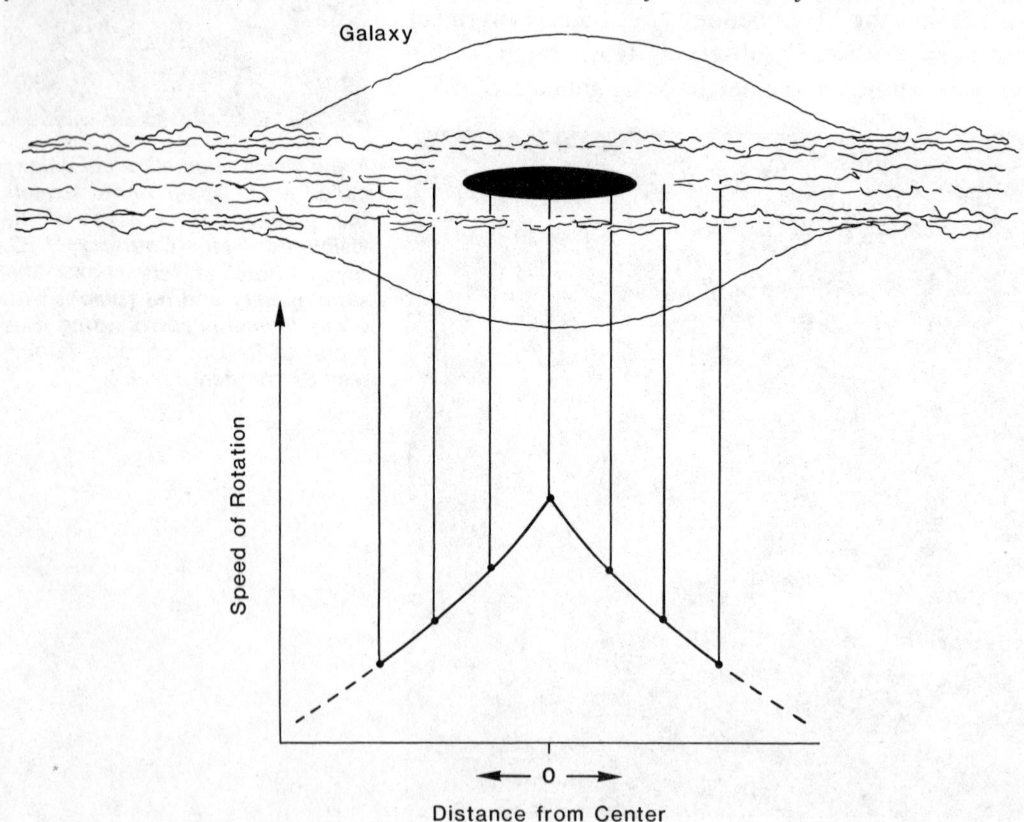

galaxies but also in their cores. The best indication of this comes from observations made by the Canadian astronomer John Kormendy in 1987. Looking at the rotation curves of the very heart of the cores of many nearby spiral galaxies, he has found that the stars in these cores are revolving at rates one can explain only through the presence of an extremely massive star or stars confined to a very small area of the heart. The calculated mass is so great, in fact, that no star or star collection could exist in so small an area without collapsing in upon itself and becoming a black hole. These galaxies are most probably homes to black holes—very massive ones, in fact—the equivalent of at least 10 million suns and invisible cosmic engines.

One final indication of dark matter comes from X-ray pictures of galactic clusters taken at the Einstein Observatory, a satellite that studied the sky from 1979 to 1983. The Einstein Observatory's results show a very hot but tenuous material surrounding many galaxies and clusters. The nature of the X-ray-emitting material is unknown, but its absence in photographs implies an M/L of greater than 600! If stars are present, they are very dim indeed.

RESOLVING THE DARK MATTER PROBLEM

In a way, the dark matter problem is a troubling challenge to astronomers—they are convinced that something exists that they can't see. This is the ultimate frustration for scientists whose work depends on observing what can be seen. The dark matter is there, and perhaps its presence can be observed through clever and indirect means. At present, we lack the ideas, if not the facilities, to pin down exactly what the form or forms of dark matter must be. What stands in the way of our fully understanding the dark matter, then, is mainly a combination of two factors: the dimness of that matter and our technological impairment in sensing it. Even in the rare cases when evidence for the dark matter can be seen with CCD arrays, it is still unclear what produces the feeble light.

The best way to resolve this problem is to get spectra of dark matter regions, especially those that form cocoons. This can show if the dark matter resembles stars or is entirely different. In this way astronomers can estimate how these stars—if the dark matter is made of stars—burn up their fuel and why they burn it in a way that prevents giving off much light. But spectral studies are very difficult to undertake at faint light levels. Will future telescopes of greater sensitivity solve the riddle of the dark matter's mysterious nature? Or will we only remain intrigued by something whose nature may not ever become known, but whose existence nonetheless dominates the makeup of the universe?

THREE

FUTURE VIEWS

· TEN ·

Frontiers—Breakthroughs and Tomorrow's Cosmology

AS of the late 1980s, cosmology stands at a crossroads between the classical cosmology, which seeks the elusive cosmic parameters through a set of continually refined standard observations, and an entirely new era of exploration, ushered in by the findings of the new cosmology. To gain a perspective on where our understanding of the universe is heading, it may be useful to recap just what we know, especially compared with what we knew ten years ago. From this we can trace the path to our most recent discoveries and tomorrow's new possibilities.

THE VIEW FROM HERE

In 1978 we knew nothing of inflation, the small fluctuations of the CBR, the structure of the DRS's, and large-scale clustering of galaxy clusters. Few astronomers felt confident that the universe held a significant amount of dark matter, that cosmic evolution should be an overriding concern in their study of the galaxies, and that the universe might not be open. Gravity lenses were little more than a theoretical construction, and distance calibration tests were emerging. Quasars were still overly mysterious objects, perhaps related to the active galaxies. Galaxies were only suspected of clustering in a structured way.

The cosmological parameters were based on broad guesswork at best; q_0 was virtually unknown. The estimated density of the universe, computed from the matter we could see, amounted to a fraction of the critical density needed to close the universe.

Today, some of these problems are still with us, but we are fast closing in on their solutions. For example, q_0 remains to be accurately estimated through tests such as the Hubble diagram. The accuracy for H_0 is still modest and may be uncertain by as much as 50%. As to the age of the universe, our figures have not changed radically in the last decade, and we still don't know if it's open or closed.

Even with these difficulties, we have a new framework, through the inflation model, for understanding why the universe came into being and for reasoning that it must be closed. New discoveries of gravity lenses, ever on the increase, will eventually allow us to probe galaxy masses and distances. Quasars have revealed themselves as distant, active galaxies, and the center of DRS's may contain black hole engines that squirt out matter in two opposing jets. Bubbles and voids of galaxy clusters appear to be a natural consequence of defects in the Big Bang, perhaps the product of cosmic strings, whose existence may yet be proved by the gravity lens effect. Finally, dark matter, irrespective of its nature, has been accepted as a major component in the overall mass content of galaxies and surrounding space.

THE NEW COSMOLOGY'S EDGE

As if to reinforce the increasing sophistication of the new cosmology, several important discoveries have come to light since 1986. Each of these key findings may prove revolutionary in our understanding of the universe. Each provides a flavor of the new cosmology's continued progress.

SNR 1987A—A Millennium Supernova

Supernovae are rare events. Perhaps one or two happen in a galaxy every century. Viewing many galaxies at once, we may, if we're lucky, spot a few supernovae every year.

Within our own galaxy the last supernova was seen over 400 years ago, so with modern instruments only the remnants of this and other nearby supernovae can be studied. But all astronomers put a new supernova on their wish list of new discoveries because supernovae reveal stars as they exist in their most extreme, if not unstable, state.

In supernovae, a red (?) giant star in a binary system actually explodes, ejecting its outer layers like a miniature Big Bang. The inner parts of the star are forced inward by the explosion, causing these parts to shrink into a very compact form; black holes may be created in this way. The expanding outer layers develop superhot temperatures and densities, causing hydrogen and helium to bind into heavier elements, the ones that make up such familiar objects as our bodies, gold nuggets, and the pages in this book.

Yet supernovae also have been used as cosmological tools, especially as distance indicators. They are also important in helping us to understand the processes by which matter reacts at very high temperatures and densities.

In February 1987, Canadian astronomer Ian Shelton discovered a new supernova in the nearby galaxy the Large Magellanic Cloud. This supernova, dubbed SNR 1987A—the first found supernova of 1987—is too close to act as a good distance calibrator for far more distant galaxies, but it is close enough to observe with a battery of telescopic instruments, providing unprecedented coverage of a supernova event. (See Plate 19 following page 142.) There have been many surprises here for the astronomer interested in the evolution of stars (for example, it turns out that the *red* giant is actually a *blue* giant). Yet one of the key values of this supernova is that it can be used to understand neutrinos, for neutrinos were discovered from this supernova's first explosive pangs.

Neutrinos rarely interact with matter, but in

certain mediums, a handful of billions may be trapped, causing slight chemical reactions inside the mediums. Neutrino telescopes are actually neutrino collectors or traps that do not give images of neutrino-emitting objects (like the sun) but rather indicate the energy and number of neutrinos that come about because of the emitter. For SNR 1987A, neutrinos were detected almost coincidentally with the first visible glimpses of the supernova, assuring that the trapped neutrinos saw a spectacular origin.

For any neutrinos to be found at all, the supernova must have produced them in almost unimaginable numbers; perhaps more than 90% of the energy of the explosion actually came out in the form of neutrinos! But are these neutrinos weightless particles, or do they have slight weights?

The weight of neutrinos is important to cosmology because, like the explosion of a supernova, the Big Bang must have generated vast quantities of neutrinos. Neutrinos do not disappear and seldom react with other matter; they are timeless Big Bang remnants. If they each have even a little weight, then their great numbers might tie up a large fraction of the mass of the universe. Neutrinos are a likely dark matter candidate; is it their total mass that closes the universe?

The best way to show if neutrinos have weight is to see how fast they move compared with the speed of light. Weightless light arrives at the observer long before particles that have weight.

Might the neutrinos from SNR 1987A have arrived shortly after the burst of light that signaled the supernova's beginning? Astronomers are trying to compare the times that neutrino events happened on that day in February and should be able to set a limit to the neutrino mass with great accuracy. The most recent analysis of the neutrinos from the neutrino telescopes does not give a tidy number to the neutrino mass, but it does set a limit of only a fraction of 1% the weight of an electron. But that limit is equally compatible with

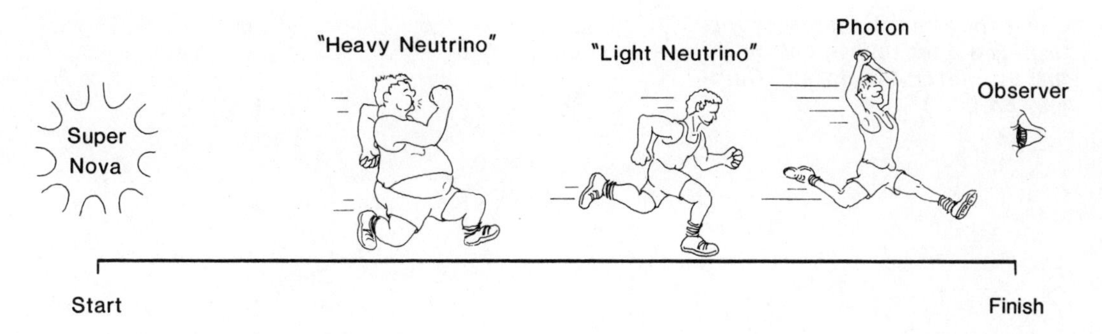

zero weight. We must wait for the weight debate to reach a final state!

Some limits on the weight of the neutrino.

Loh and Spillar's Test

The source counts test is plagued with problems, including that of galactic evolution. Yet if we could undertake a cosmological test that was not dependent on galactic evolution, it should be possible to find the value of q_0.

Princeton physicists Edwin Loh and Earl Spillar have done just that by devising a new version of the source counts test that gives an unambiguous estimate of q_0. Their test, now called the Loh and Spillar comoving density test, differs from other methods in that it uses galaxies as markers of the volume of the universe in increments of redshift, rather than of magnitude. There should be slight differences in the number of galaxies in a redshift increment from nearby redshift (distances) to distant redshift. But the method is very sensitive to small changes in the volume attributed to the braking of the universe (or, synonymously, its curvature) and can show differences in the volume of space for two relatively nearby redshift increments.

This test does not take into account the actual magnitude of galaxies—only their numbers and redshifts—and in this way it is similar to the source counts tests devised by Sandage (see Chapter 3). But Loh and Spillar also observed thousands of galaxies, in several slices of sky, to get enough data

Using their novel source counts test, Loh and Spillar calculated that the universe is closed. (Courtesy Ed Loh.)

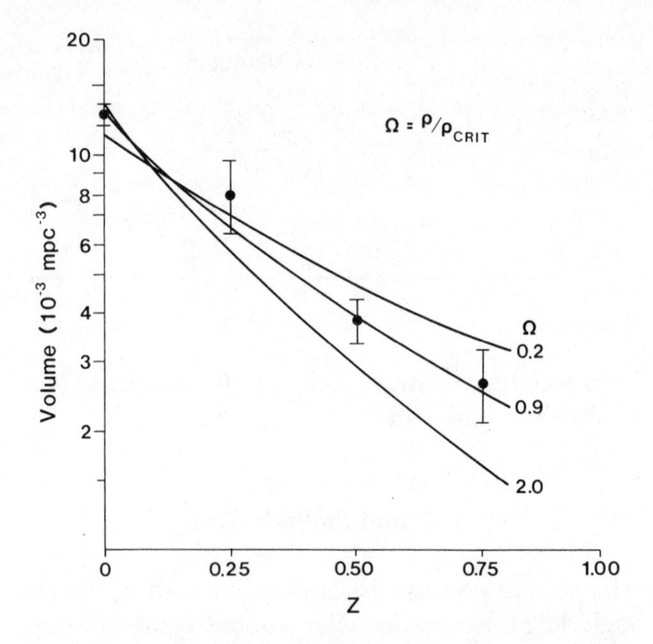

to make the test meaningful. They combined a new approach of analysis with a more thorough sample of observations than any previously taken.

The neatest aspect of this test is this: Not only does it function without relying on the magnitudes of the galaxies; it even functions without relying on their *M/L* ratio. So the galaxies are merely markers for the volume of space at different points in the universe's expansion, not its mass. The test requires no distinction between dark matter and the matter we see.

As of 1986, results of the test are quite favorable to the inflation theory. The value of q_0 is within 10% to 20% that needed to close the universe; the density of the universe today is at or near the critical density. In this one test, Loh and Spillar show that the universe may be closed. Thus the dark matter must be prevalent to make up the remaining 85% of the mass needed to close the universe.

If we may presume that the universe is closed, then suddenly, from the Friedmann equation, it becomes clear that Hubble's constant should lie between 40 and 60 km/s/megaparsec. Here is an incentive to use the distance indicators to get H_0 with greater accuracy and check the conclusion of Loh and Spillar.

Dim Galaxies and the Angular Size-Redshift Test

American physicist Anthony Tyson, a pioneer in CCD array use in astronomy, has used CCD to make observations of galaxies dozens of times dimmer than those ever seen before. This has proved several interesting points: (1) the galaxies actually do start increasing in angular size at large redshifts, and (2) the quasars themselves form clusters just like other galaxies. These two new findings may eventually enable us to use the angular size-redshift test to check the value of q_0 or, if Loh and Spillar are correct, to find out more precisely how galaxies evolve and how galaxies show the cluster structure.

Other Universes (?)

We speak of there being one universe. This may not be so, if the latest theories about GUTS (see Chapter 6) are to be believed.

In GUTS, physicists find that the easiest way to unite the forces of nature is to make the dimensions manyfold, even more so than our usual notions of up-down, right-left, toward-away, and of time. Now, within a theory called *superstrings* (not to be confused with cosmic strings), GUTS predicts that there are 10 dimensions; the remaining 6 are rolled up into a string. This idea is called *compactification.* We never notice the other seven dimensions; they are effectively dormant to our universe.

And now for the fantastic part—unification can break down at any time and at any place, causing the other dimensions to unfurl and form a new

If "existence" comprises 10 dimensions, the remaining 6 must be compactified. This may be analogous to forcing a three-dimensional space to two dimensions or to one dimension.

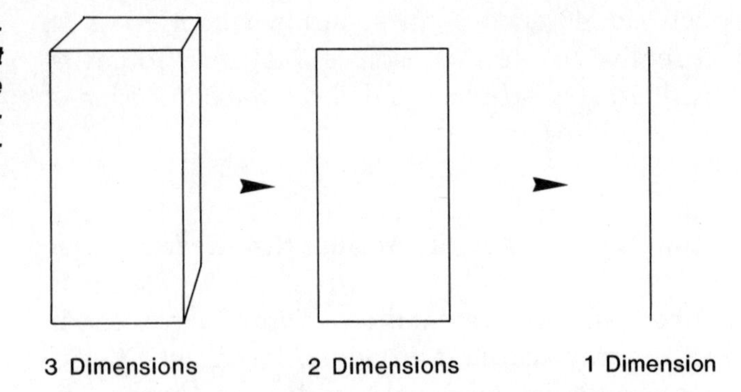

combination. So if we live in dimensions 1, 2, 3, 4, another set may come into being, made up of 1, 2, 3, 9, for example. This is an entirely different universe, budded off, or cloned, from another one. It is independent of the old universe and has its own set of physical laws; it is a totally unknown form of reality. Contact between universes is impossible. Each is an end unto itself.

In 1987, Soviet physicist Andrei Linde showed that the idea of *clone universes* was consistent with the inflationary model of the Big Bang. Indeed, it is in extreme conditions, as in inflation, that cloning is possible.

Where might the conditions for universe cloning also occur? Alan Guth has considered this question and found that black holes might be a good source. Perhaps each new black hole gives rise to a new universe, as unrolling dimensions, unfurled in the black hole's near-infinity, come into being. And with this idea, profound philosophical and theological questions come to mind: What caused what? Is something behind this strange cosmic order? If something cannot be *detected* in any way, does it really *exist* in nature? If we go to high enough densities, can we make a universe in a laboratory? These far-out speculations are not beyond the present conjecture of some physicists—and ultimately they may prove to be substantiated. But right now they are intellectual constructions of great profundity waiting for observations and experiments to bring them to light.

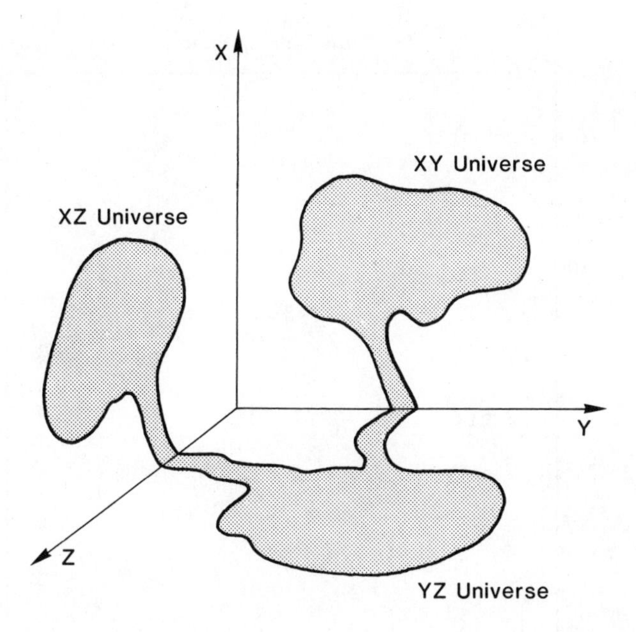

Clone universes, with different dimensionalities.

A Possible Cosmic String

Cosmic strings (see Chapters 6 and 8), the defects at the Big Bang's beginning, have seemed elusive to astronomers so far. Yet an odd view of some galaxies, in 1987, has raised the question of whether a cosmic string has been found.

In an observation made by the Canadian astronomer Lennox Cowie and the American astronomer Esther Hu, a CCD image revealed a patch of sky in which at least four pairs of galaxies are present. Each pair had members separated by about two arcseconds, and each member has about the same redshift as the other. Yet between pairs, the redshifts were quite different.

The easiest way to explain this strange grouping of galaxy pairs is through a combination of gravity lensing and cosmic strings. Here, a massive cosmic string lies between us and several distant galaxies. As the light from each one of these galaxies passes the string, the string acts as a bender, forming two images of each galaxy. The fact that each pair shows the same separation between members is

These four galaxy pairs show up in a small region of sky. Are they gravity-lens images of a foreground cosmic string? (Courtesy L. Cowie and E. Hu.)

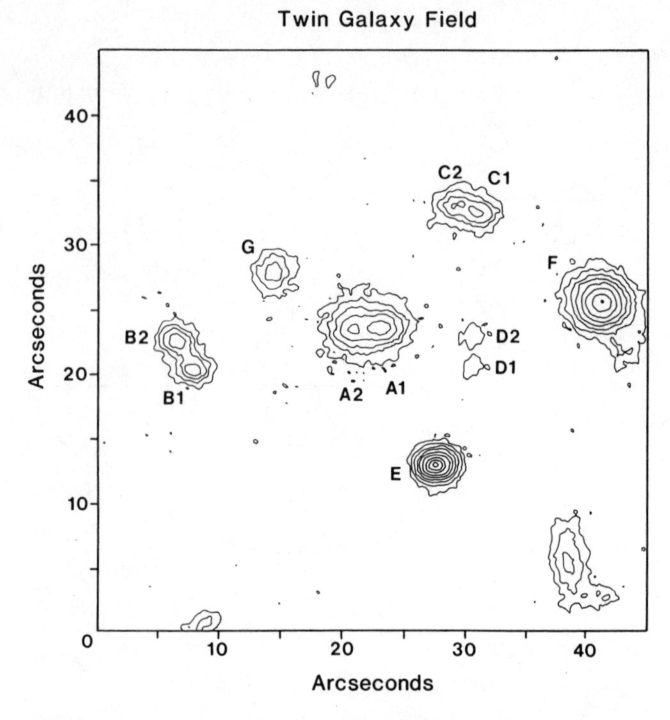

very improbable for any kind of bender except a cosmic string.

We have a choice here—either a cosmic string has lensed several background galaxies, or we are coincidentally looking at a patch of sky where several binary galaxies exist. Extraordinary hypotheses, such as a cosmic string, require extraordinary evidence, so Cowie and Hu are continuing their observations to get better spectra of the galaxies and to get better hints of each member's shape and size. If the spectra and shapes in each pair turn out

A cosmic string reveals itself through generating double images of background objects.

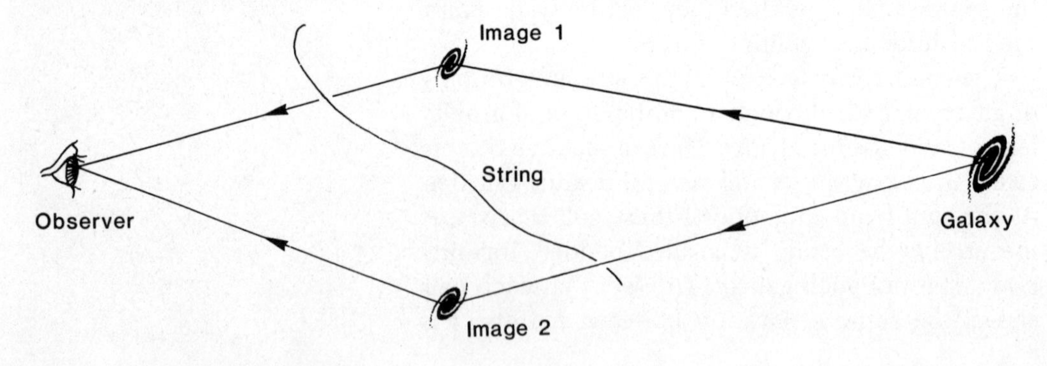

to be exact, then a cosmic string will have been found, clinching the notion of the inflationary universe, where strings must have been made. Then we need to see if more cosmic strings abound, perhaps as attractors of galaxies, perhaps as the hearts of galaxies themselves.

Cosmic Arcs—Gravity's Megalens

Views of galaxies and clusters always reveal surprising shapes and structures. But once every few years, something very different shows up in photographs or CCD images, defying conventional explanations.

The latest in these surprises are the cosmic arcs. These are narrow, arched strips of light found in the vicinity of some galactic clusters. They are often quite faint, and only a handful have been found so far.

What makes the cosmic arcs so unusual is that

The cosmic arc 2242-02, a gravity-lens crescent where a cluster is the bender. (Courtesy NOAO.)

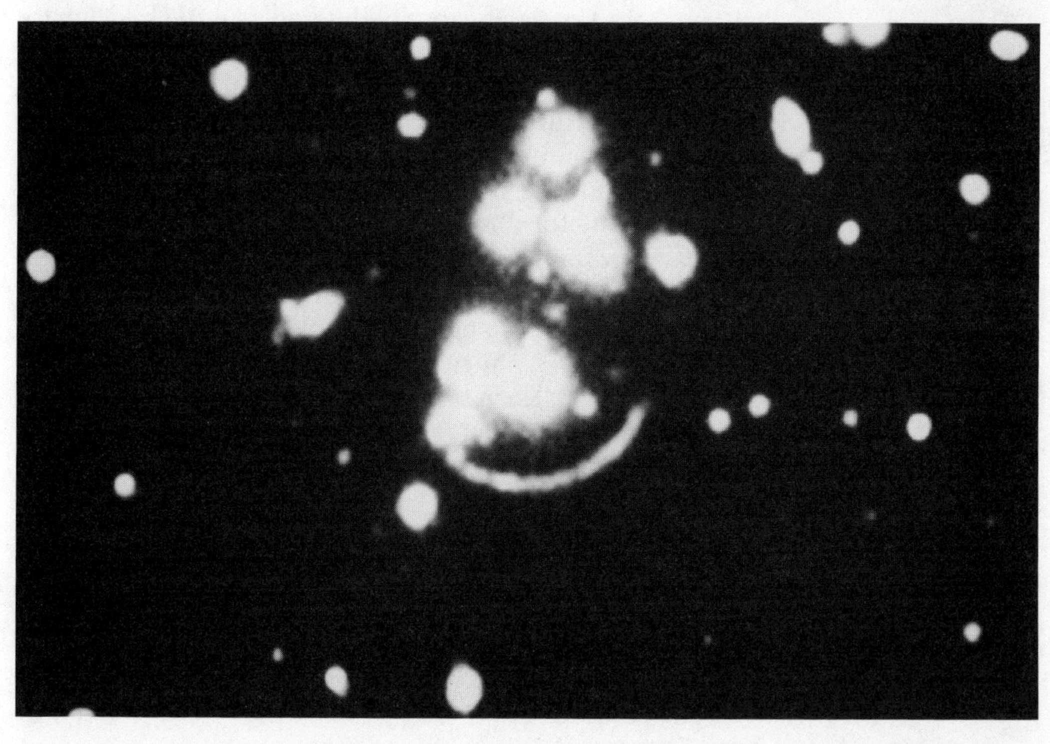

they are not part of galaxies but arise in the vicinity of clusters, with apparent sizes as big as the clusters themselves. And they are always ringlike, with the cluster located at the ring's center.

If one were to imagine the gravitational lens effect as a likely candidate for explaining the arcs, one might be right. But the problem with these arcs is that they are huge—many arcminutes across—whereas the image separations found in other gravity lenses is only a few arcseconds. In 1987, however, Bernard Fort, a French astronomer, and independently American astronomers Roger Lynds and Vahe Petrosian observed one cosmic arc, associated with the cluster called Abell 370. They found that the arc had a redshift that was identical over its length and very different from the redshift of the cluster itself. These points were convincing evidence that Abell 370, an entire cluster, acts as a bender in forming a gravity lens mirage—part of an Einstein ring—of a distant galaxy. (See Plate 28 following page 142.)

The significance of this finding is twofold. First, it places a limit on the mass of a galactic cluster that is at least 100 times higher than other estimates; the cluster must be 10^{15} times more massive than the sun. Second, it is an absolute clincher for the existence, if not the form, of dark matter; the *M/L* ratio for this cluster may be greater than 1,000! Almost none of the mass in Abell 370 emits light. This cosmic arc has born the fruit of gravity lenses as a way of probing the universe and revealing hints of its mass. Other megalenses are being sought right now to learn something about distant galaxies otherwise hidden from our view.

GREAT LIGHT HOPES—THE NEXT TELESCOPES AND NEW EYES

Earth is a pleasant place to live, but it has its problems as a site for astronomical observations. Ranking highest on the list of these problems is light pol-

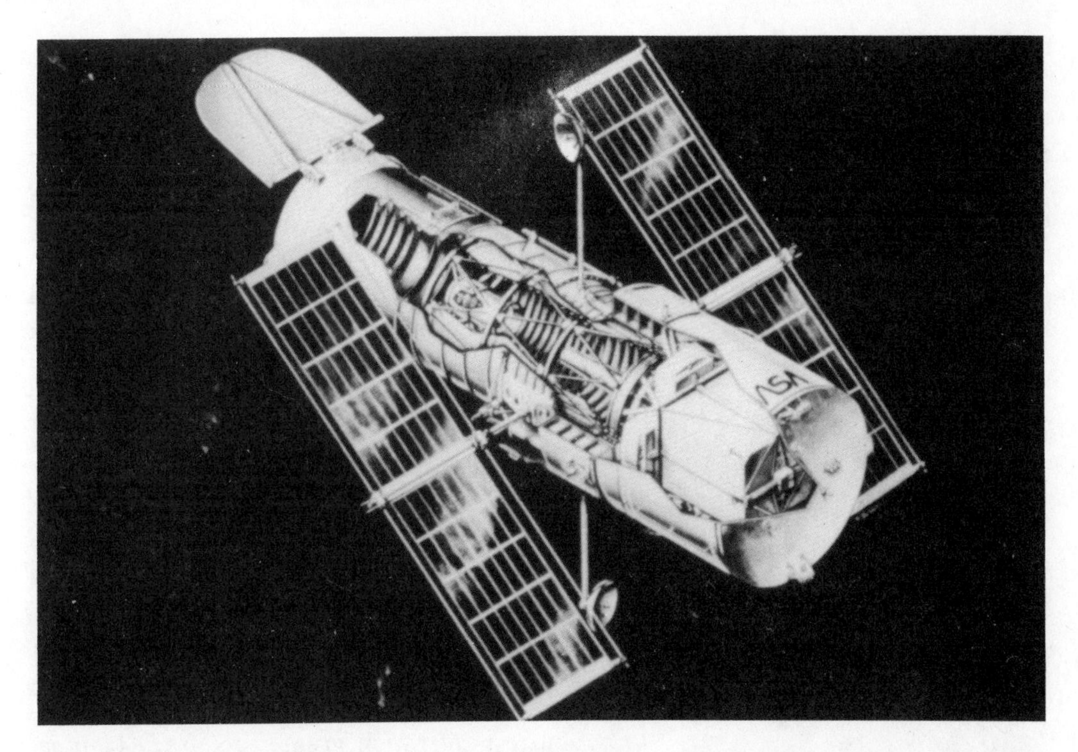

The Hubble Space Telescope, which, when launched, will be a valuable tool in cosmology. (Courtesy NASA.)

lution and the twinkling effect of the atmosphere. Also, the atmosphere is opaque to some wavelengths, such as X-rays and far infrared. Above the blanket of air, space-based telescopes suffer from none of these problems.

The greatest advantage to telescopes in space is their freedom from the atmosphere and dedication to viewing a few objects at a time. They can get much higher resolution and sensitivity, revealing features that are invisible from the ground. Galaxies more than 100 times fainter than those we're now straining to see might be detected, and the wispy cocoons and galactic clusters might be more readily apparent.

Astronomers have long dreamed of an optical telescope in space, and now the culmination of those efforts has brought the Hubble Space Telescope (HST) into being. Patiently waiting in a warehouse until its pre-1990 launch date, the HST is appropriately named, for many of its biggest chores relate to cosmological tests first used by Hubble himself. Here are a few of the things that the HST will be able to do from space:

1. Find dim Cepheids in galaxies over a dozen times more distant than those that are now known, lending observations of much greater distance precision to the Hubble diagram.
2. Detect the outer structures of galaxies and show the spectra and distribution of dark matter in the cocoons.
3. Show how the angular size of galaxies changes at great distances.
4. Measure the very dim spectral lines from distant, cold gas clouds to get the relative abundance of deuterium in them (this should corroborate predictions on the amount expected from the Big Bang's first minutes of being).
5. Reveal how the bubbles and voids at very great distances are structured.
6. Provide measurements for all the cosmological tests at much greater accuracy of measurement.
7. Study the streaming motion of galaxies in clusters at distances many times more distant than those seen at present.
8. Find out how much the effects of galactic evolution are important to the farthest galaxies, so that evolution effects can be separated from those of the expanding universe.

Although the HST will be the first effective optical telescope in space, telescopes probing radio, infrared, ultraviolet, and x-rays have been part of the satellite-borne approach to astronomy for years. So far, two of these satellites, the Einstein Observatory, active in the late 1970s, and the IRAS satellite, active in the mid 1980s, have made modest contributions to cosmology.

The Einstein Observatory revealed an x-ray view of the sky, the first glimpse of the universe with a sensitive x-ray eye. Quasars provided the most

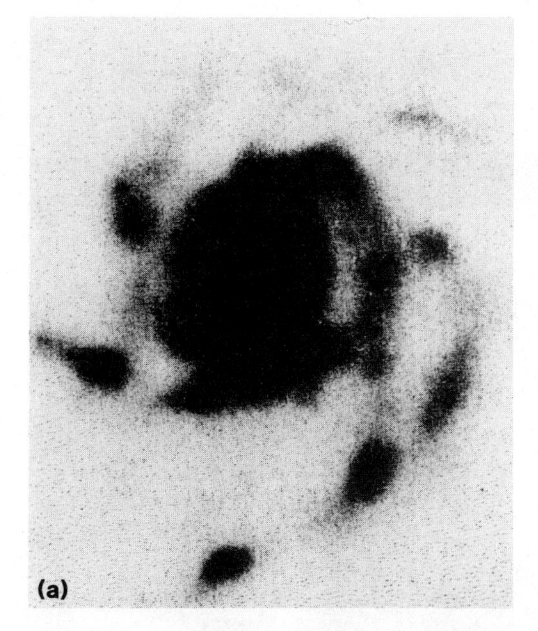

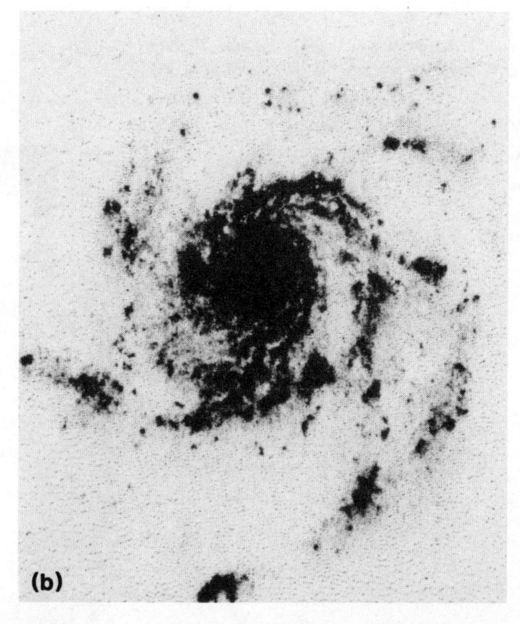

The same galaxy viewed from Earth (a) and simulated clarity expected with HST (b). (Courtesy John L. Tonry.)

curious results from these x-rays. They were found to be prodigious emitters of x-radiation, countless times stronger than normal galaxies. Like some galaxies, quasars may be enveloped in a cocoon of hot, x-ray-emitting gas, perhaps part of the complex environment of the earlier universe.

IRAS, an infrared telescope, was a unique and continuing success story that allows the infrared sky to be mapped precisely and sensitively. Ironically, so much data has been gathered from IRAS that it may take a decade or more to understand it all. Eventually, the source count test can be tried from this infrared data, which may be less hampered by the galactic evolution effects seen at visible wavelengths.

Curiously, much of tomorrow's observing needs to be done from the ground. No matter how much better space is for freedom from pollution and atmosphere, there is no substitute for bigger telescopes of greater sensitivity. Huge telescopes are now being planned to probe the deepest edge of the galactic realm.

Bigger telescopes gather more light and thus

A future ground-based telescope of tremendous size and light-gathering power will be made of novel mirror designs, including multiple mirrors of small size. (Courtesy Roger Angel.)

see dimmer objects better. This is crucial to spectroscopy, which has always been plagued by the limits of telescopic size in acquiring enough light to make spectra. Because spectral lines are so narrow in wavelength, spectroscopy is never as sensitive to dimness as a photographic or electronic image. Then again, spectroscopy is primarily concerned with the total brightness of an entire object at different wavelengths and so is unaffected by twinkling or smearing. And because spectra tell

This odd structure is actually a mirror furnace for baking telescope mirrors eight meters across. (Courtesy Roger Angel.)

much about the redshift, distance, and conditions in galaxies and quasars, they are essential in the study of the farthest cosmic realms.

Making huge mirrors, such as those of the five-meter Palomar Telescope in California or the six-meter Special Astrophysical Observatory in the Soviet Union, is a fragile and frustrating challenge. Large mirrors crack or deform from their own weight. It was assumed for many years that these two would be the largest telescopes ever built. However, recent technology has afforded new techniques for forming, polishing, and mounting very large mirrors. Super-thin mirrors, only a few centimeters thick in spots, get by these problems. British-American astronomer Roger Angel, and others at Caltech and Chile, are devising ways of constructing very large mirrors, some eight meters or more in size.

One particularly promising remedy to large mir-

Multiple mirror designs combine many mirrors to gather the light sensitivity found in one large one. Spectroscopy, in particular, can be done most effectively with large mirror and multiple mirror designs.

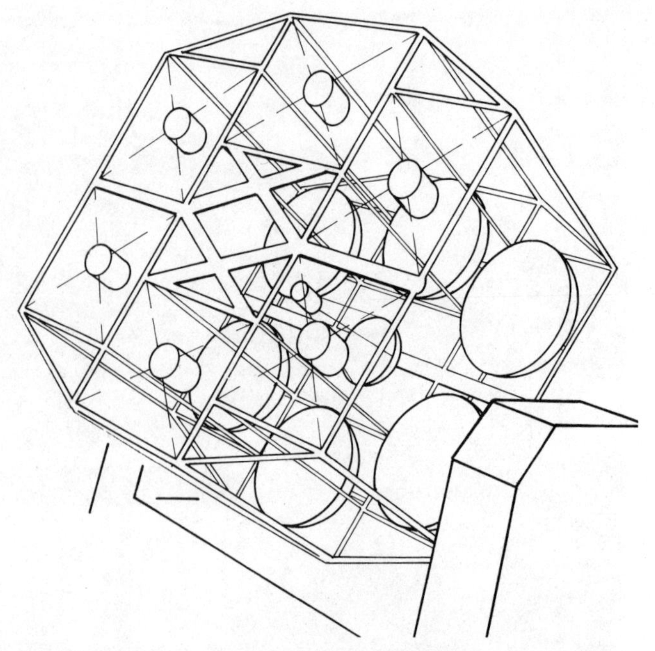

ror woes is to combine the light reflected from many smaller mirrors. These are called multiple mirror designs. The reason it took so long to try this technology (the first multiple mirror design, the MMT in Arizona, has been utilized only for a decade) is the headache of keeping many mirrors precisely pointed at one spot for long periods of time. But with computers, it is possible to keep mirrors aligned precisely and instantaneously. In the future, almost all new telescopes will incorporate multiple mirror designs. With these, as with all large single telescopes, spectroscopy can be done effectively, probing new territory and perhaps clarifying questions about dark matter, protogalaxies, and more. Future views will be exciting ones.

· ELEVEN ·

Cosmological Conclusions

AS a field of human endeavor, cosmology can boast that it has deciphered the most out of the least. Few other disciplines are as restrictive in terms of what, how much, and what kind of data can be gathered for study. But despite this, cosmologists have progressed from naked eye to electronic image, from dogmatic belief to rigorous analytical framework. And this progress, with the last decade's new cosmology, has been rapid and enlightening.

Our present understanding of the universe is based on the Big Bang, for its predictions have been tested, through observation, and found to be consistent with prediction. Yet it is the gross attributes of the Big Bang model—the CBR, the recession of galaxies, the formation of the lightest elements—that are its strongest points. What has emerged from relative simplicity in theory fits what we see. But the Big Bang, as a theory, is only two generations old, a mere point in time compared to the millennium of incorrect explanations of the Greeks and others. The Big Bang will stay around because it works best and will remain the theory of choice until, if ever, some inexplicable new discovery contradicts its predictions, making the Big Bang untenable or vastly incomplete.

There have, however, been radical improvements in our observations of the universe; it is here that the change has been most marked. Small modifications and gains in sensitivity give big payoffs. Even views of well-studied objects yield uncanny details to old knowledge. Time and again, it is the improvements of factors of two or three—a telescope just a little bigger, a method a little less noisy—rather than those improvements of factors of ten or more, that open windows to new discoveries. There can never be an ultimate telescope or electronic eye. Each new piece of data is a critical addition to the grasp of cosmic knowledge.

If the pace of progress has been rapid in the last decade, it has come solely as the result of taking risks—lightly. Ignorance is a powerful fuel for its own eradication, and every astronomer assumes, at some level, that he or she is ignorant of some aspect of their study. If astronomers knew, beforehand, exactly what they would find, many evenings spent in frigid darkness could be passed up for the warmth of a well-lit office. They would have no need to observe. Sometimes, the observer's payoff is serendipitous and important, other times it is apparently inconsequential. But diamonds do not appear in every pile of volcanic grit. They must be sought out at the price of patience and persistence.

It is this persistence that makes cosmology both an evolving science and an incomplete one. Many of the universe's details have been, and remain, unknown to us. Yet a marked number of important details are accessible to us. How many of the unknowns are ultimately inponderable? Are there limits to what we can ever know about nature? Will we devise new methods of viewing presently hidden vistas? Can we assume that limits are not synonymous with our intellectual limitations? Cosmology needs to include such questions, just as it considers the value of Hubble's constant.

We end this summary of our present cosmic knowledge on a better footing than we began. Yet we live in a changing, growing universe. We should not be too surprised if our knowledge continues to change with it, even as we stand on the edge of understanding the treasure within our grasp.

Glossary

Accretion Disk

The ring, or pancake-like structure, of superhot gas that forms around a black hole or other compact star.

Active Galaxy

A galaxy with a core whose magnitude changes radically, indicating violent activity from a cosmic engine.

Angular Momentum

A quantity related to the speed and mass of a rotating object or orbiting point. A galaxy with a great deal of angular momentum will be turning rapidly (every 100 million years) compared to one with little angular momentum.

Angular Size

How much of the sky an object subtends, expressed as an angle, usually in units of arcseconds.

Angular Size-Redshift Test

A cosmological test that compares the angular size of galaxies at different redshifts to look for any curvature in the universe.

Arcsecond

A unit of angle, equal to 1/3600 of a degree.

Bender

A mass, usually a galaxy or a cluster of galaxies, that bends the light from a background galaxy or quasar to form the gravity lens mirage of the background object.

Big Bang

A model, now accepted as a standard, that the universe had a beginning as an incredibly compact point that has been expanding ever since.

Black Hole

A star that has collapsed onto itself, creating densities that are so high that matter cannot exist in normal forms. The shrinking of the star's radius has built up a velocity of escape that is so high that light cannot escape—hence the name black hole.

Brown Dwarf

A star about one-fifth the mass of the sun that is not quite hot enough to start its nuclear fires. It is a feeble emitter of infrared radiation and is nearly invisible.

Bubble

One name for the structuring of galaxy and galaxy clusters around a void.

CBR

Cosmic background radiation—a continuous glow of radiation that can be seen in the microwave and infrared part of the spectrum. The CBR arose in the early universe where light and matter decouple; light shines through what was previously an opaque universe. The CBR stage occurred before the formation of galaxies.

CCD

Stands for charge coupled device, an array of semiconductors that are sensitive to light. They register light and transmit it as a current, which can be used with computers to record an image.

Cepheid

A variable star whose time scale or period of variation is linked with its absolute magnitude; a Cepheid can be used to infer distances via the distance modulus.

Closed Universe

The state when the universe stops its expansion and begins to contract.

Compactification

The squeezing together of the other six dimensions that make up space; we live in a universe where only a specific four dimensions are important.

Core

The inner heart of a galaxy, its brightest and most massive part.

Cosmological Tests

Methods used to get the cosmological quantities from observed properties of galaxies.

Critical Density

The density at which the universe becomes closed, stopping its expansion, as an onset to a contraction stage.

Curvature

The slight but universal bending of space in the universe caused by the gravity of its mass.

Dark Matter

The mass of the universe, or of specific galaxies, that does not give off any light and is thus invisible.

Deceleration Rate

The slowing down of the Big Bang's expansion, usually symbolized by q_0. It is closely related to the universe's curvature; $q_0 = 0$ in a flat, ever-expanding universe.

Distance Calibrator

A way of getting distances to galaxies, which then allows an estimate of H_0.

Distance Modulus

A method of getting distances, which results from a comparison of measured (what you see) magnitude and absolute (intrinsic) magnitude.

Doppler Effect

An object's wavelength shift caused by a receding or forward speed to the observer.

DRS

Double radio source—the dumbbell-shaped structure of radio galaxies and quasars that comprises a core, jets, and lobes.

Einstein Ring

When an observer, bender, and background object are perfectly aligned in a gravity lens, the background object appears as a mirage shaped like a ring. The angular size of the ring can be used to infer the mass of the bender.

Engine

Or cosmic engine—the "something," probably a black hole, responsible for the activity in the core of galaxies, especially the DRS phenomenon and quasars.

Evolution

The change in a galaxy's characteristics, such as absolute magnitude, over time.

Expansion Rate

Another name for Hubble's constant, H_0. It is usually expressed as velocity per distance increment and preferred units are km/s/megaparsec. H_0 is the slope of the line given by Hubble's law.

Flatness Problem

The problem of explaining how the early universe flattened out all or most of its curvature.

Geodesic

The path or line that light rays take; the shortest path between us and the radiating source.

Globular Cluster

A spherical star cluster that may contain over 100,000 stars. Important in finding the size of the Milky Way and the age of the universe.

Gravitational Lens Effect

The name for the arrangement of an observer, bender, and background object that gives the mirage of the background object.

Great Attractor

An unknown, massive object or objects, whose gravitational attraction causes the streaming motion of our galaxy and many others nearby.

GUTS

Grand Unified Field Theories—the general class of ideas that account for the situation in which the different forces are interchangeable.

Horizon Problem

The problem in explaining the great isotropy, or smoothness of the CBR.

Hubble Flow

The velocities that correspond to the expansion of the universe at various distances.

Hubble's Law

The relation between galaxy distance and its redshift, indicating the expansion of the universe.

Inflation

A model for the earliest moments of the Big Bang where the forces become un-unified and the universe inflates during the resulting phase transition.

Interferometry

A method of attaining high resolution by connecting telescopes spanning great distances; the largest distances are achieved with Very Long Baseline Interferometry (VLBI).

Inverse Square Law

The way that light drops off in intensity with increasing distance. It is expressed as $I = I_0/D^2$ where I = the measured intensity, I_0 = the initial intensity, and D = the distance.

Isotropy

A description of the universe that says that it looks the same in every direction.

Jet

The funneled squirting of hot electron gas from an engine to a lobe in a DRS.

Jetstar

Miniature versions of DRS's; binary stars within our galaxy.

Large-Scale Structure

General term for the connective pattern of galaxy clusters across hundreds or more megaparsecs.

Lyman α (alpha) Forest

The plethora of absorption spectral lines seen in the direction of distant quasars. It may indicate gas clouds left over, or forming protogalaxies in a dense environment.

Magnitude

A logarithmic scale of brightness where each unit of magnitude equals a 2.512 brightness difference. Absolute magnitude refers to brightness at a fixed or reference distance; measured magnitude is the brightness we actually see.

Mergers

The coming together or collision of galaxies to form distorted new ones.

M/L

The mass-to-light ratio—a way of describing an object's mass according to how much light it gives off. Large *M/L* objects are called dark matter. The sun has an *M/L* of 1, by definition.

Neutrino

An elementary particle that seldom reacts with regular matter. Neutrinos were formed in the earliest Big Bang and may be around today, making up a large amount of the dark matter.

Normal Galaxy

A galaxy whose core maintains a steady magnitude over time.

Nucleosynthesis

The making of heavier elements from lighter ones, such as the combination of hydrogen atoms to form helium. Most nucleosynthesis happens in the cores of stars or in supernova explosions.

Open Universe

The case where the universe continues to expand because there is insufficient matter to stop the expansion.

Parallax

The proportional shift in an object's angular position based on a change of perspective by the observer.

Parsec

A unit of distance derived from parallax but used to describe all cosmic distance scales. A megaparsec is a million parsecs; a kiloparsec is a thousand parsecs.

Quark

The tiniest of elementary particles, present in the earliest Big Bang.

Quasars
Active galaxies that inhabit the early universe.

Redshift
The Doppler shift of an object as it moves away, or recedes, from us.

Rotation Curve
A graph of the velocity of rotation for points at increasing distances from a galaxy's core.

Source Counts Test
A cosmological test that infers the universe's density, or curvature, by comparing the number of galaxies found at different redshifts or magnitudes.

Spectrum
Breaking up of radiation into component wavelengths.

Streaming
Motion of galaxies that is large, oriented toward a specific direction, and unrelated to the Hubble flow.

Strings
Or cosmic strings—energetic defects in the Big Bang. Strings act like skinny but long masses.

Strong Force
One of the four forces of nature; it is important only over small distances (a fraction of an atom's size).

Superstrings
A theory that the "stuff" of matter consists of 10 dimensions.

Tully-Fisher Relation
A distance calibrator method that infers a galaxy's absolute magnitude from its 21-centimeter velocity width.

Unification
The joining of the four forces of nature during the moments before the inflationary stage of the Big Bang.

Virial Theorem
A way to estimate masses of galaxies based on the assumption that their energy of motion and their gravitational energy have balanced off.

Void

A hole in space, absent of galaxies, many hundreds, if not thousands, of megaparsecs across.

Weak Force

Another force of nature, not quite as strong in intensity as the strong force.

Z

A way of describing redshift that relates to the wavelength shift caused by the Doppler effect and corrects for the special relativity effects of large velocities.

21-Centimeter Line

A radio spectral line of unionized hydrogen used to trace the gas in galaxies.

Index